Worked Examples in Modern Physics

VOLUME 2

Worked Examples in Modern Physics

VOLUME 2

P. ROGERS, M.SC., and G. A. STEPHENS, Ph.D.

LONDON ILIFFE BOOKS LTD

First published in 1967 *by*
Iliffe Books Ltd., Dorset House
*Stamford Street, London, S.E.*1

Printed in England at the Pitman Press, Bath

Publishers' Note

The publishers would like to thank the Governing Bodies of the Universities for permission to include in this book examples from their examination papers and at the same time to state that these Bodies accept no responsibility for the accuracy or content of the solutions given, nor are they in any way committed to approval of the answers.

Preface

The answering of numerical questions plays an important part in all undergraduate physics courses, and success in this field may well lead to a more incisive approach to other physics problems and, of course, to improved paper qualifications.

The book is intended primarily for undergraduates reading for an ordinary degree or its equivalent, and particularly for those part-time students who have insufficient access to their tutors. Nevertheless, the honours student will find much of interest to him.

The aim of the book is to give its readers confidence in tackling numerical questions in a wide spectrum of topics in modern physics. The problems have been compiled from examination papers set recently at various universities and colleges, and are grouped under the following main headings: Atomic and Electron Physics and Nuclear Physics in Volume 1; X-rays and Crystal Structure, Solid State Physics, Wave Mechanics, and the Special Theory of Relativity in Volume 2.

The comprehensive list of contents should enable the reader to find easily the chosen topic. A representative selection of problems in each section have been answered in full; the remainder being left to the student. It is suggested that the student should first attempt the problems without referring to the worked solutions or answers. In this way the difficulties of the problem will be fully appreciated.

Although the questions were chosen primarily for their numerical content, the descriptive part of the questions will enable the student to realize the extent of his knowledge of the subjects concerned and will be invaluable in the preparation for examinations.

P. R.
G. A. S.

September, 1966

Contents

3. X-RAYS AND CRYSTAL STRUCTURE

Origin of X-Rays **127**

3.1 Minimum wavelength of the continuous spectrum 127
3.2 Characteristic X-ray spectra and Moseley's law 128
3.3 Minimum voltage to excite characteristic spectra 129
3.4 Relationship between absorption edge and excitation voltage 130
Unworked problems 131

Absorption of X-Rays **132**

3.9 Absorption of X-rays in an alloy 132
3.10 Relationships between absorption coefficients 134
Unworked problems 135

Bragg's Law **135**

3.13 Simple example of Laue technique 135
3.14 Determination of Bragg angles 137
3.15 Effective wavelength of electrons 138
3.16 Relationship between lattice parameters, molecular weight and density 139
3.17 Determination of the coefficient of thermal expansion using X-rays 140
Unworked problems 141

Missing Orders **144**

3.26 Diffraction of thermal neutrons by α-iron (b.c.c.) 144
3.27 Position of first diffraction maximum in the powder diffraction photograph of a b.c.c. material 145

3.28 Analysis of powder diffraction photograph of a b.c.c. material 146
3.29 Analysis of powder diffraction photograph of a f.c.c. material 148
Unworked problem 149

Miscellaneous **150**

3.31 Diffraction of X-rays by a ruled grating 150
3.32 Effect of grain size on a powder diffraction photograph 151

4. SOLID STATE PHYSICS

Lattice Energy of Ionic Crystals **153**

4.1 Energies due to the Coulomb attraction of neighbouring ions and repulsion between the electron shells 153
4.2 Compressibility of an ionic salt 155

Magnetic Properties of Materials **157**

4.3 Determination of the magnetic dipole moment of Ni^{++} from the Quincke experiment 157
4.4 Bohr magneton value of nickel 159
4.5 Saturation intensity of nickel 159
4.6 Bohr magneton value of a mixed ferrite 160
4.7 Adiabatic demagnetisation of gadolinium sulphate 162
Unworked problem 164

Point Defects **165**

4.9 Density of Schottky defects 165
Unworked problem 166

The Free Electron and Zone Theories **166**

4.11 Calculation of Fermi energy of sodium 166
4.12 Maximum kinetic energy of electrons in the first zone of a simple cubic material 167
Unworked problem 168

Free Charge Carriers in Metals and Semiconductors **169**

4.14 Relationship between charge carrier density, mobility and electrical conductivity 169

4.15 Relaxation time of free electrons 170
4.16 Simple Hall coefficient 171
4.17 Interpretation of the Hall coefficient 171
4.18 The effect of *n*- and *p*-type impurities in germanium 172
Unworked problems 174

Miscellaneous **175**

4.22 Thermionic emission 175
4.23 Edge dislocations at a low angle grain boundary 176

5. WAVE MECHANICS

Dualism **178**

5.1 *and* 5.2 The Compton scattering of X-rays 178
5.3 The de Broglie wavelength of a particle 180
5.4 The refractive index of a metal for electrons 181
5.5 *and* 5.6 Electron diffraction 182
Unworked problems 185

Heisenberg Uncertainty Principle **187**

5.11 Energy of a particle in an infinitely high potential well 187
5.12 Total energy of a harmonic oscillator 187
Unworked problem 188

The Schrödinger Wave Equation **189**

5.14 Reflection of electrons at a potential step 189
5.15 Transmission of particles through a rectangular potential barrier of finite length 191
5.16 The bound states of a particle in a square well 196
Unworked problems 201

Operator Formalism **202**

5.19 Expectation values for a linear harmonic oscillator 202
5.20 Ground state of the hydrogen atom 205
Unworked problem 206

6. SPECIAL THEORY OF RELATIVITY

Michelson-Morley Experiment **208**

6.1 Calculation of expected fringe shift 208

The Lorentz Transformation **210**

6.2 Addition of velocities—decay of a radioactive nucleus 210
6.3 Relative velocities between colinear particles 212
6.4 Relative velocities between coplanar particles 213
6.5 Elastic scattering of relativistic particles 216
6.6 Time and space dilation 220
6.7 Decay of the μ-meson produced by cosmic rays 221
6.8 Relativistic Doppler shift 223
Unworked problems 225

Fundamental Physical Constants and Symbols

FUNDAMENTAL PHYSICAL CONSTANTS

The following values have been used unless otherwise specified

Term	Symbol	Value	
Avogadro's number	$\mathbf{N_0}$	$6{\cdot}025 \times 10^{23}$	mole^{-1}
Boltzmann's constant	$\mathbf{k}$	$1{\cdot}38 \times 10^{-16}$	erg/deg
		$1{\cdot}38 \times 10^{-23}$	J/deg
Planck's constant	$\mathbf{h}$	$6{\cdot}62 \times 10^{-27}$	erg sec
		$6{\cdot}62 \times 10^{-34}$	J sec
Velocity of light	$\mathbf{c}$	$3{\cdot}0 \times 10^{10}$	cm/sec
Electronic rest mass	$\mathbf{m}_e$, $\mathbf{m}_0$	$9{\cdot}11 \times 10^{-28}$	g
Electronic charge	$\mathbf{e}$	$1{\cdot}60 \times 10^{-20}$	e.m.u.
		$1{\cdot}60 \times 10^{-19}$	C
		$4{\cdot}80 \times 10^{-10}$	e.s.u.
Specific charge on an electron	$\mathbf{e}/\mathbf{m}_e$	$1{\cdot}76 \times 10^{7}$	e.m.u./g
		$1{\cdot}76 \times 10^{11}$	C/kg
Atomic mass unit (a.m.u.)		$1{\cdot}66 \times 10^{-24}$	g
Mass of hydrogen ion	$\mathbf{m}_{\mathrm{H}}$	$1{\cdot}66 \times 10^{-27}$	kg
		931 erg/gauss	MeV
1 electron volt	eV	$1{\cdot}60 \times 10^{-12}$	erg
		$1{\cdot}60 \times 10^{-19}$	J
Bohr magneton	μ_{B}	$0{\cdot}927 \times 10^{-20}$	erg/gauss
		$9{\cdot}27 \times 10^{-24}$	$\mathrm{Am^2}$
Permittivity of free space	ε_0	$8{\cdot}85 \times 10^{-12}$	F/m
Permeability of free space	μ_0	$4\pi \times 10^{-7}$	H/m

SYMBOLS

The following symbols imply the following quantities unless otherwise specified

Symbol	Term
A	Atomic number
a	Lattice parameter
B	Flux density
$\mathbf{c}$	Velocity of light
d	Lattice spacing
E	Electric Field Strength
$\mathbf{e}$	Charge on a particle
$\mathbf{g}$	Acceleration due to gravity
g	Lande splitting factor
$\mathbf{h}$	Planck's constant
j, J	Total angular momentum
$\mathbf{k}$	Boltzmann's constant
k	Wave number ($2\pi/\lambda$)
l, L	Orbital angular momentum
M	Molecular weight
m	Mass of an electron (or particle concerned)
m^*	Effective mass of electron
$m^*(X)$	Effective mass of electron in an atom 'X'
$\mathbf{m}_e$, $\mathbf{m}_0$	Rest mass of an electron
$\mathbf{m}_{\mathrm{H}}$, $\mathbf{m}_{\mathrm{P}}$	Mass of hydrogen ion (proton)
m_l, m_J	Magnetic quantum number
n	Principal quantum number
p	Momentum

Symbol	Term
Q	Energy released in nuclear reactions
R, r	Radius of curvature
t, τ	Time
T	Absolute temperature or Radioactive half-life
v	Velocity
V	Potential difference
Z	Atomic Number
γ	$(1 - v^2/c^2)^{-\frac{1}{2}}$
η	Viscosity
λ	Wavelength or Radioactive Decay Constant
μ	Absorption coefficient or Mobility
μ_B	Bohr magneton
ν	Frequency
$\bar{\nu}$	Wave number $(1/\lambda)$
ρ	Density
σ	Conductivity or Cross section
χ	Mass Susceptibility

3
X-rays and Crystal Structure

ORIGIN OF X-RAYS

Problem 3.1

Describe how X-ray wavelengths have been measured. Derive all formulae used.

If the minimum wavelength recorded in the continuous X-ray spectrum from a 50 kV tube is 0·247 Å, calculate the value of Planck's constant.

[B.Sc. Applied Science, Pt. I, Durham, 1963]

SOLUTION

The energy, E, gained by each electron as it passes along the X-ray tube equals the electronic charge, **e**, multiplied by the potential difference, V, through which the electrons fall, i.e.

$$E = \mathbf{e}V = 1{\cdot}6 \times 10^{-19} \times 50 \times 10^{3} \text{ joule}$$

For an X-ray of minimum wavelength, λ_{min}, (i.e. maximum energy) all this energy must be converted into one X-ray quantum. The energy of such an X-ray quantum

$$\mathbf{h}\nu_{max} = \frac{\mathbf{hc}}{\lambda_{min}} = \mathbf{e}V$$

where **h** is Planck's constant, i.e.

$$\mathbf{h} = \frac{\mathbf{e}V\lambda_{min}}{\mathbf{c}} = \frac{1{\cdot}6 \times 10^{-19} \times 50 \times 10^{3} \times 0{\cdot}247 \times 10^{-10}}{3 \times 10^{8}}$$

$$= \underline{6{\cdot}58 \times 10^{-34} \text{ joule sec}}$$

Problem 3.2

Write an account of (*a*) the continuous X-ray spectrum, and (*b*) the characteristic X-ray spectrum. Give an interpretation of the spectra in terms of the theory of the nuclear atom.

Show how **h**/**e**, where **h** is Planck's constant and **e** is the electronic charge, may be determined from experimental data on the continuous X-ray spectrum.

X-rays from a tube with a cobalt target show strong *K*-lines for cobalt ($Z = 27$) and weak *K*-lines due to impurities. The wavelengths of the K_α-lines are 1·785 Å for cobalt and 2·285 Å and 1·537 Å for the impurities. Using Moseley's law, calculate the atomic number of each of the two impurities, taking the 'screening constant' for the *K*-lines as unity.

[Dip. Tech. Pt. I, Salford, 1961]

SOLUTION

By Moseley's law, the frequency ν, of a particular characteristic X-ray line (e.g. K_α) of various materials is given by

$$\nu = C(Z - a)^2$$

where C is a constant, Z is the atomic number of the target and a is the 'screening constant'. For the K_α-lines

$$\nu = C_{K_\alpha}(Z - 1)^2$$

where C_{K_α} is the appropriate value of C for the K_α-lines, or working in terms of wavelengths

$$\lambda = \frac{\mathbf{c}}{\nu} = \frac{C'_{K_\alpha}}{(Z-1)^2}$$

where C'_{K_α} is another constant. For cobalt $Z = 27$ and $\lambda = 1{\cdot}785$ Å. Therefore

$$C'_{K_\alpha} = 1{\cdot}785 \times 26^2 \text{ Å}$$

The wavelength due to the first impurity $\lambda_1 = 2{\cdot}285$ Å thus its atomic number Z_1 is given by

$$Z_1 - 1 = \left(\frac{C'_{K_\alpha}}{\lambda_1}\right)^{\frac{1}{2}} = \left(\frac{1{\cdot}785}{2{\cdot}285}\right)^{\frac{1}{2}} \times 26 = 23$$

$\therefore \qquad Z_1 = 24$, which corresponds to chromium

And similarly for the second impurity

$$Z_2 - 1 = \left(\frac{1{\cdot}785}{1{\cdot}537}\right)^{\frac{1}{2}} \times 26 = 28$$

$\therefore$ $\underline{Z_2 = 29}$, which corresponds to copper

Problem 3.3

Discuss the spectrum of X-rays emitted from a target bombarded by fast electrons.

Calculate the minimum voltage required to produce a certain K_α line of wavelength 0·5 Å.

[B.Sc. (Normal), Pt. I, Liverpool, 1963]

SOLUTION

To produce a K_α line an electron must be removed from the K-shell ($n = 1$), and according to theory this requires an energy

$$E_K = \frac{2\pi^2 m e^4 Z_e^2}{\mathbf{h}^2}$$

where Z_e is the effective atomic number of the atom for that electron. The K_α-line is produced due to an electron transition from the L shell ($n = 2$) to the K-shell, and the corresponding energy change is

$$\Delta E = \frac{2\pi^2 m e^4 Z_e^2}{\mathbf{h}^2}\left(\frac{1}{1^2} - \frac{1}{2^2}\right)$$

and substituting for E_K we get

$$\Delta E = \tfrac{3}{4}E_K$$

Thus in terms of wavelengths, the line is described by

$$\lambda_{K\alpha} = \frac{\mathbf{c}}{\nu} = \frac{\mathbf{ch}}{\Delta E} = \frac{4\mathbf{ch}}{3E_K}$$

The ionisation energy E_K is then given by

$$E_K = \frac{4\mathbf{ch}}{3\lambda_{K\alpha}}$$

and substitution yields

$$E_K = \frac{4 \times 3 \times 10^{10} \times 6{\cdot}62 \times 10^{-27}}{3 \times 0{\cdot}5 \times 10^{-8}} = 5{\cdot}296 \times 10^{-8} \text{ erg}$$

$$= \frac{5{\cdot}296 \times 10^{-8}}{1{\cdot}60 \times 10^{-12}} = 3{\cdot}31 \times 10^{4} \text{ eV}$$

The potential to excite the K_α line is thus $\underline{33{\cdot}1 \text{ kV}}$

Problem 3.4

Show that the quantum theory predicts one *K*- and three *L*-absorption edges in a metal, and describe briefly the photo-electric experiments by Robinson which showed this to be so.

Calculate the voltages required to excite the K_α and K_β lines of cobalt.

$$\lambda_{K_\alpha} = 1{\cdot}787 \text{ Å (mean)}$$
$$\lambda_{K_\beta} = 1{\cdot}617 \text{ Å}$$
$$\lambda_K = 1{\cdot}604 \text{ Å (absorption edge)}$$

[B.Sc. (Special), Sheffield, 1954]

SOLUTION

To excite the *K*-series of X-ray lines, electrons must be removed from the *K*-shell, and this requires energy

$$E_K = \frac{hc}{\lambda_K}$$

where λ_K is the wavelength of the absorption edge. Therefore

$$E_K = \frac{6{\cdot}62 \times 10^{-27} \times 3 \times 10^{10}}{1{\cdot}604 \times 10^{-8}} = 1{\cdot}238 \times 10^{-8} \text{ erg}$$

$$= \frac{1{\cdot}238 \times 10^{-8}}{1{\cdot}604 \times 10^{-12}} = 7{\cdot}74 \times 10^{3} \text{ eV}$$

∴ The potential to excite K_α and K_β lines is $\underline{7{\cdot}74 \text{ kV}}$.

Problem 3.5

Discuss briefly the design and operation of gas filled and high vacuum X-ray tubes.

In an X-ray tube, the electrons are accelerated by a potential difference of 50 kV. Calculate the wavelength of the X-rays produced assuming half the energy of each electron is converted into an X-ray quantum.

[Dip. Tech. Elec. Eng., Salford, 1959]

Ans. 0·497 Å

Problem 3.6

An X-ray tube has an anode of tin ($Z = 50$). What is the minimum voltage which must be applied to the tube in order to produce the K-lines? Derive any formulae used, using Bohr's simple quantum theory.

[B.Sc. (Normal), Pt. II, III, Liverpool, 1961]

Ans. 32·7 kV

Problem 3.7

Calculate the wavelength of the X-ray line emitted when an electron falls from the L-Shell to the K-Shell of a Fermium atom ($Z = 100$). Use Bohr's simple quantum theory to derive any formulae used.

[B.Sc. (Normal), Pt. II, III, Liverpool, 1962]

Ans. 0·124 Å

Problem 3.8

Give a brief account of the emission and absorption X-ray spectra of an element and show in detail how they can be explained in terms of the structure of the atom.

The wavelengths of the absorption edges of molybdenum are

$K = 0{\cdot}620$ Å; $L_{\mathrm{I}} = 4{\cdot}299$ Å; $L_{\mathrm{II}} = 4{\cdot}722$ Å; $L_{\mathrm{III}} = 4{\cdot}914$ Å.

Calculate the wavelengths of the $K_{\alpha 1}$- and $K_{\alpha 2}$-lines in the X-ray spectrum of molybdenum and show the transitions involved on an energy level diagram.

[Dip. Tech. Pt. I, Salford, 1962]

Ans. $\lambda_{K_{\alpha 1}}$ 0·710 Å; $\lambda_{K_{\alpha 2}}$ 0·714 Å

ABSORPTION OF X-RAYS

Problem 3.9

Describe the changes which occur in the X-rays emitted by an X-ray tube with a copper target as the potential difference across the tube is changed from 5 kV to 30 kV, and explain these changes.

Explain fully how you would obtain monochromatic X-rays from such a tube. Under what conditions would this radiation be unsuitable for obtaining a powder pattern of a substance and why?

X-rays from a tube with a copper target pass through a piece of foil made of an alloy of 30% copper and 70% nickel by weight. If the thickness of the foil is $2{\cdot}00 \times 10^{-3}$ cm, what is the ratio of the intensities of the K_α and K_β components of the transmitted beam? Ratio of intensities of K_α to K_β components in the unfiltered beam $= 5{\cdot}00:1$

Mass absorption coefficient of copper for Cu K_α radiation $= 52{\cdot}9\ \text{cm}^2\ \text{g}^{-1}$
Mass absorption coefficient of copper for Cu K_β radiation $= 39{\cdot}3\ \text{cm}^2\ \text{g}^{-1}$
Mass absorption coefficient of nickel for Cu K_α radiation $= 45{\cdot}7\ \text{cm}^2\ \text{g}^{-1}$
Mass absorption coefficient of nickel for Cu K_β radiation $= 275{\cdot}0\ \text{cm}^2\ \text{g}^{-1}$
The density of the alloy $= 8{\cdot}91\ \text{g cm}^{-3}$

[H.N.C. Applied Physics, Salford, 1963]

SOLUTION

If an X-ray beam of intensity I_0 is made to pass through m g per unit area of a material of mass absorption coefficient μ_m then the emerging intensity is

$$I = I_0 \exp[-\mu_m m] \tag{3.9.1}$$

If the X-ray beam passes through masses of two materials the emerging intensity is

$$I = I_0 \exp - [\mu_{m1}m_1 + \mu_{m2}m_2] \qquad (3.9.2)$$

where the subscripts 1 and 2 refer to the respective materials.

The density of the absorbing foil $\rho = 8{\cdot}91$ g/cm³. This is equivalent to an area density ρ_A given by

$$\rho_A = \rho t$$

where t is the thickness of the foil. Thus

$$\rho_A = 8{\cdot}91 \times 2 \times 10^{-3} = 1{\cdot}782 \times 10^{-2}\ \text{g/cm}^2$$

The area density of copper in the foil

$$= \frac{1{\cdot}782 \times 10^{-2} \times 30}{100} = 0{\cdot}5346 \times 10^{-2}\ \text{g/cm}^2$$

The area density of nickel in the foil

$$= \frac{1{\cdot}782 \times 10^{-2} \times 70}{100} = 1{\cdot}247 \times 10^{-2}\ \text{g/cm}^2$$

The intensity of the K_α-line I_α, after passing through the foil is, from Equation 3.9.2 above, given by

$$I_\alpha = I_{\alpha 0} \exp - [(0{\cdot}5346 \times 52{\cdot}9 + 1{\cdot}247 \times 45{\cdot}7)10^{-2}]$$

where $I_{\alpha 0}$ is its initial intensity. Therefore

$$I_\alpha = I_{\alpha 0} \exp [-0{\cdot}8527]$$

or

$$\frac{I_\alpha}{I_{\alpha 0}} = 0{\cdot}4263$$

Similarly for the K_β-line

$$I_\beta = I_{\beta 0} \exp - [(0{\cdot}5346 \times 39{\cdot}3 + 1{\cdot}247 \times 275)10^{-2}]$$
$$= I_{\beta 0} \exp [-3{\cdot}64]$$

or

$$\frac{I_\beta}{I_{\beta 0}} = 0{\cdot}02627$$

$$\therefore \qquad \frac{I_\alpha}{I_\beta} = \frac{I_{\alpha 0}}{I_{\beta 0}} \times \frac{0{\cdot}4263}{0{\cdot}02627} = \frac{5 \times 0{\cdot}4263}{0{\cdot}02627} = \underline{81}$$

Problem 3.10

(*a*) What are the hazards to health which may occur in X-ray laboratories? What methods are adopted to control them?

(*b*) Define the terms linear-absorption coefficient, mass-absorption coefficient and gram-atomic-absorption coefficient and explain the conditions under which they are valid.

(*c*) Calculate the linear absorption coefficient of sodium nitrite ($NaNO_2$) for copper K_α radiation given the following data:

Density of sodium nitrite = $2{\cdot}16$ g cm^{-3}

Element	*Atomic weight*	*g. atomic absorption coefficient* cm^2 (g. atom)$^{-1}$
sodium	23·00	711
nitrogen	14·01	119
oxygen	16·00	203

[H.N.C. Applied Physics, Salford, 1962]

SOLUTION

The gram molecular-absorption coefficient $\mu_{\text{gm. mol.}}$ is given by

$$\mu_{\text{gm. mol.}} = \mu_m M$$

where μ_m is the mass absorption coefficient and M is the molecular weight of the substance.

$\mu_{\text{gm. mol.}}$ may also be expressed in terms of the gram-atomic-absorption coefficients ($\mu_{\text{gm. at.}}$) of the constituent atoms of the molecule thus:

$$\mu_{\text{gm. mol.}}(\text{NaNO}_2) = \mu_{\text{gm. at.}}(\text{Na}) + \mu_{\text{gm. at.}}(\text{N}) + 2\mu_{\text{gm. at.}}(\text{O})$$

$$\therefore \qquad \mu_m M = \mu_{\text{gm. mol.}}(\text{NaNO}_2) = 711 + 119 + 203 \times 2$$

$$= 1{,}236 \text{ cm}^2 \text{ mole}^{-1}$$

$$\therefore \qquad \mu_m = \frac{1{,}236}{23 + 14 + 32} = 17{\cdot}92 \text{ cm}^2/\text{g}$$

But the linear absorption coefficient $\mu = \rho\mu_m$ where ρ is the density of the material. Therefore

$$\mu = 17{\cdot}92 \times 2{\cdot}16 = \underline{38{\cdot}7 \text{ cm}^{-1}}$$

Problem 3.11

Describe briefly the ways in which a material reduces the intensity of an X-ray beam.

A narrow beam of X-rays containing radiations of two wavelengths is incident on an absorber for which the linear absorption coefficients at the two wavelengths are 3·0 cm^{-1} and 10 cm^{-1} respectively. If at incidence the ratio of the intensities of the two radiations is unity, for what thickness of absorber will the ratio of the intensities be 2 to 1?

[B.Sc. (Subsidiary), Bristol, 1963]

Ans. 0·099 cm

Problem 3.12

Discuss briefly the biological dangers of ionising radiations and explain the meaning of the terms roentgen, rad, r.b.e., and rem.

Describe an ionisation chamber suitable for measuring X-ray beam intensities.

A parallel beam of monochromatic X-rays enters such a chamber and its intensity is found to be 1,000 units. Calculate the intensities measured when 2 mm thick sheets of aluminium and copper are, in turn, placed in the beam. What is the half-value thickness of aluminium for this wavelength?

Density of aluminium = 2·7 g cm^{-3}
Density of copper = 8·93 g cm^{-3}
Mass absorption coefficient of aluminium = 0·143 cm^2 g^{-1}
Mass absorption coefficient of copper = 0·263 cm^2 g^{-1}

[Dip. Tech. Elec. Eng., Salford, 1961]

Ans. Al, 926 units; Cu, 714 units, 1·79 cm

BRAGG'S LAW

Problem 3.13

Describe and explain the methods of observing the X-ray diffraction patterns of single crystals.

A transmission Laue photograph is made with a cubic crystal of lattice parameter 4·00 Å.

Calculate the minimum distance from the centre of the pattern of reflections from the (111) planes, if the specimen is 5 cm from the film and the X-ray tube voltage is 40 kV.

[Dip. Tech. Pt. I, Salford, 1963]

SOLUTION

The Bragg relation states $n\lambda = 2d \sin \theta$, and so for θ_{min} we need the minimum wavelength available and a first order reflection, i.e.

$$\sin \theta_{\text{min}} = \frac{\lambda_{\text{min}}}{2d}$$

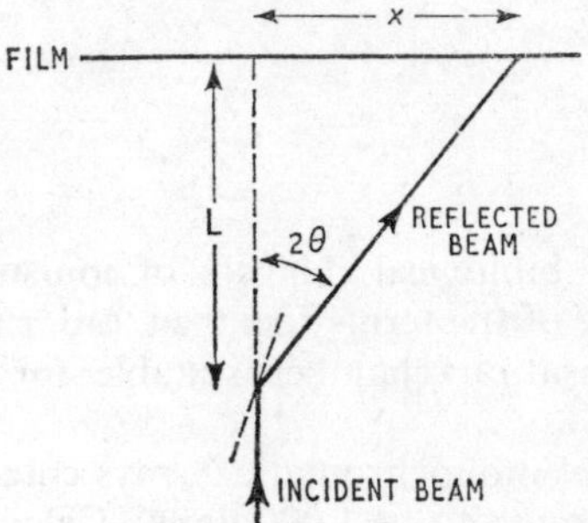

Fig. 3.1

From the geometry of cubic crystals the spacing of the (111) planes is

$$d = \frac{a}{(h^2 + k^2 + l^2)^{\frac{1}{2}}} = \frac{a}{\sqrt{3}}$$

where a is the lattice parameter and h, k and l are the Miller indices of the plane concerned and λ_{min} is given as in Problem 3.1 by

$$\lambda_{\text{min}} = \frac{\mathbf{hc}}{\mathbf{e}V}$$

where V is the tube voltage. Therefore

$$\sin \theta_{\text{min}} = \frac{\mathbf{hc}\sqrt{3}}{2\mathbf{e}Va}$$

$$= \frac{6{\cdot}62 \times 10^{-34} \times 3 \times 10^{8} \times \sqrt{3}}{2 \times 1{\cdot}6 \times 10^{-19} \times 40 \times 10^{3} \times 4 \times 10^{-8}} = 0{\cdot}06717$$

and $$\theta_{\text{min}} = 3^{\circ}\, 51'$$

From the geometry of the Laue arrangement, as shown in Fig. 3.1,

$$x_{\min} = L \tan 2\theta_{\min}$$

since the angle of deviation of the X-ray beam is twice the Bragg angle. Substituting, we get

$$x_{\min} = 5 \tan 7^\circ 42' = \underline{0{\cdot}676 \text{ cm}}$$

Problem 3.14

Explain what is meant by the terms 'unit cell' and 'Miller indices'. Calculate the smallest and the largest possible Bragg angle for the reflection of X-rays ($\lambda = 1{\cdot}436$ Å) from a simple cubic crystal, if the edge of the unit cell is 3·607 Å. You may assume the relation $2d \sin \theta = n\lambda$, but should prove all other equations used.

[B.Sc. (Special), Pt. I, Bristol, 1961]

SOLUTION

From the Bragg equation, $n\lambda = 2d \sin \theta$, and for $\theta_{\max}$

$$\sin \theta_{\max} = \frac{n\lambda}{2d}$$

should approach unity as nearly as possible.

Now from the crystal geometry the crystal spacing

$$d = \frac{a}{\sqrt{(h^2 + k^2 + l^2)}}$$

where d is the edge of the unit cell, i.e. $d = a/\sqrt{N}$, where for a simple cubic $N = 1, 2, 3, 4, 5$ etc., but not 7, 15, or 23. Therefore

$$\sin \theta = \frac{n\lambda\sqrt{N}}{2a} = \frac{1{\cdot}436}{2 \times 3{\cdot}607} n\sqrt{N} = 0{\cdot}1991\, n\sqrt{N}$$

In the *first* order $n = 1$, therefore, $\sin \theta = 0{\cdot}1991\sqrt{N}$ and the largest possible value of N is 25, giving $\sin \theta = 0{\cdot}9955$.

In the *second* order $n = 2$, $\sin \theta = 0{\cdot}3982\sqrt{N}$ and the largest possible value of N is 6, giving $\sin \theta = 0{\cdot}977$.

In the *third* order $n = 3$, $\sin \theta = 0{\cdot}5973\sqrt{N}$ and the largest possible value of N is 2, giving $\sin \theta = 0{\cdot}845$.

In the *fourth* order $n = 4$, $\sin\theta = 0{\cdot}7964\sqrt{N}$ and the largest possible value of N is 1, giving $\sin\theta = 0{\cdot}7964$.

In the *fifth* order $n = 5$, and if we have $N = 1$ we get the same maximum value of $\sin\theta$ as in the first order, i.e. $\sin\theta = 0{\cdot}9955$.

Higher orders are not possible. Therefore, θ_{max} is given in the first order, for $N = h^2 + k^2 + l^2 = 25$, or the fifth order for $N = 1$. Thus

$$\theta_{max} = \sin^{-1} 0{\cdot}9955 = \underline{84° \ 30'}$$

For θ_{min}, $n = 1$ and $d = d_{max} = a$. Therefore

$$\sin\theta_{min} = \frac{\lambda}{2a} = \frac{1{\cdot}436}{2 \times 3{\cdot}607} = 0{\cdot}1991$$

and

$$\theta_{min} = \underline{11° \ 29'}$$

Problem 3.15

How can a monochromatic beam of X-rays be produced? Explain fully the principles of the methods you suggest.

What is the wavelength of a beam of X-rays which is diffracted by a crystal in a direction making an angle of 60° with the incident beam and corresponding to first-order 'reflection' from crystal lattice planes separated by 3×10^{-8} cm? What would be the velocity of a beam of electrons, diffracted in the same direction by the same crystal?

[B.Sc. (Subsidiary), Bristol, 1960]

SOLUTION

The Bragg angle is half the angle through which the beam is deviated, thus in this case the Bragg angle, $\theta = \frac{1}{2} \times 60° = 30°$.

In the first order, from the Bragg equation, we have

$$\lambda = 2d \sin\theta = 2 \times 3 \times \sin 30° = 3 \text{ Å}$$

Electrons of the same effective wavelength would be diffracted similarly. The effective wavelength λ_e of such electrons in terms of their velocity v is, by the de Broglie relation, given by

$$\lambda_e = \frac{\mathbf{h}}{mv}$$

$$\text{or} \quad v = \frac{\mathbf{h}}{m\lambda_e} = \frac{6{\cdot}62 \times 10^{-27}}{9{\cdot}11 \times 10^{-28} \times 3 \times 10^{-8}} = \underline{2{\cdot}42 \times 10^8 \text{ cm/sec}}$$

Problem 3.16

Show how Bragg's Law for reflection of X-rays by crystal planes can be derived from the Laue conditions for diffraction from the points in a 3-dimensional crystal lattice.

A narrow pencil of monochromatic Cu K_α X-radiation is allowed to fall on a crystal of silver bromide (face-centred cubic). A first order reflected beam from a (111) plane is observed at an angle of 26° 40′ to the incident beam. Calculate the wavelength of the X-radiation employed.

Density of silver bromide $= 6{\cdot}47$ g cm^{-3}
Molecular weight of silver bromide $= 188$

[B.Sc. (Special), Pt. I, Bristol, 1960]

SOLUTION

As in the previous examples, the crystal plane spacing is given by

$$d = a/\sqrt{h^2 + k^2 + l^2}$$

Thus

$$d_{(111)} = \frac{a}{\sqrt{(1^2 + 1^2 + 1^2)}} = \frac{a}{\sqrt{3}}$$

By the Bragg relation, for first order reflections

$$\lambda = 2d \sin\theta = \frac{2a}{\sqrt{3}} \sin\frac{26^\circ\ 40'}{2}$$

Thus

$$\lambda = 0{\cdot}2663a \qquad (3.16.1)$$

In the unit cell of the f.c.c. crystal there are eight corner silver atoms each contributing $\frac{1}{8}$ atom to the cell, since each is shared by eight similar cells. Also there are six face-centred silver atoms each contributing $\frac{1}{2}$ atom to the unit cell since each is shared by two cells. The effective number of silver atoms per unit cell is then, $8 \times \frac{1}{8} + 6$

$\times \frac{1}{2} = 4$. Thus there will also be 4 bromine atoms associated with each unit cell.

$$\text{The density of silver bromide} = \frac{\text{mass of unit cell}}{\text{volume of unit cell}}$$

hence

$$6{\cdot}47 = \frac{4 \times 188 \times \mathbf{m}_H}{a^3}$$

but

$$m_H = \frac{1}{N_0}$$

Therefore

$$a^3 = \frac{4 \times 188}{6{\cdot}47 \times 6{\cdot}025 \times 10^{23}}\ \text{cm}^3$$

giving $a = 5{\cdot}78$ Å. From Equation 3.16.1 above we have

$$\lambda = 0{\cdot}2663 \times 5{\cdot}78 = \underline{1{\cdot}54\ \text{Å}}$$

Problem 3.17

Describe and explain how the X-ray diffraction patterns of polycrystalline materials can be observed, and indicate the applications of the method.

When X-rays are incident on a polycrystalline material at 24°C, a diffracted beam is observed at 164·5° to the incident beam, and when the material is at 170°C the diffracted beam is at 162° to the incident beam. Calculate the coefficient of linear expansion of the material and indicate how the data could be obtained experimentally.

[H.N.C. Applied Physics, Salford, 1962]

SOLUTION

The Bragg relation states $n\lambda = 2d \sin \theta$. Since at the two temperatures we are considering the same line and we are using the same monochromatic X-ray beam then $n\lambda$ is constant and

$$d_1 \sin \theta_1 = d_2 \sin \theta_2$$

where d_1, d_2, θ_1 and θ_2 are the crystal spacings and diffraction angles at the two temperatures. Hence

$$\frac{d_2}{d_1} = \frac{\sin \theta_1}{\sin \theta_2}$$

giving

$$\frac{d_2 - d_1}{d_1} = \frac{\sin \theta_1 - \sin \theta_2}{\sin \theta_2}$$

The coefficient of linear expansion $\alpha = (d_2 - d_1)/d_1 t$ where t is the corresponding temperature difference. Therefore

$$\alpha = \frac{\sin \theta_1 - \sin \theta_2}{t \sin \theta_2}$$

The Bragg angles θ_1 and θ_2 are, however half the angles through which the incident beam is deviated. Therefore

$$\alpha = \frac{\sin 82{\cdot}25^\circ - \sin 81^\circ}{146 \sin 81^\circ} = \frac{0{\cdot}0031}{146 \times 0{\cdot}9877} = \underline{2{\cdot}15 \times 10^{-5}\,{}^\circ\mathrm{C}^{-1}}$$

Problem 3.18

Describe the powder camera used with X-rays and show how it can be used to study the crystal structure of sodium chloride.

A crystal has atoms equally spaced along the X, Y and Z directions, the separation of the atoms along these directions being 3 Å. A beam of white X-rays is incident on the crystal along the X direction and a beam is observed emerging along the Y direction. What is the wavelength of this diffracted beam?

[Dip. Tech. Elec. Eng., Salford, 1963]

Ans. $3/n$ Å, $n = 1, 2, 3, \ldots$

Problem 3.19

Determine the cube edge and fraction of the volume occupied by hard spheres of radius R in a body-centred cubic lattice.

The wavelength of a particular X-ray line from a molybdenum target is 0·71 Å. The line is incident on a NaCl crystal. Calculate the first-order reflection angles from (100), (110) and (111) planes, ($d_{100} = 2{\cdot}84$ Å).

Account for the fact that the first-order Bragg reflection from the

(111) planes of NaCl is quite strong but is almost entirely missing for the analogous case of KCl.

[B.Sc. (General Hons.), Durham, 1962]

Ans. θ_{100}, 7° 11′
θ_{110}, 10° 11′
θ_{111}, 12° 30′

Problem 3.20

Explain how X-rays are diffracted by a single crystal and describe the single crystal X-ray spectrometer.

Monochromatic X-rays fall on a single crystal of calcite. Two successive angles to the surface at which diffraction takes place are 11° 32′ and 17° 28′. If the grating spacing of the calcite is 3·035 $\times 10^{-8}$ cm determine the wavelength of the X-rays and the orders of the diffractions occurring.

[B.Sc. (Normal), Pt. II, III, Liverpool, 1962]

Ans. 0·607 Å
Orders 2 and 3

Problem 3.21

Potassium chloride (KCl) is a cubic crystal and has a density of 1·98 g cm^{-3}. What is the distance between neighbouring atoms? Compute all the angles for which Bragg reflections can take place from X-rays of wavelength 2·0 Å. Explain why this number is finite—even when all crystal planes are considered.

(At. wt. of K = 39·10, At. wt. of Cl = 35·45.)

[B.Sc. (Special), Pt. I, Bristol, 1962]

Ans. 3·15 Å, 18° 30′, 26° 30′,
33° 18′, 39° 21′, 45° 6′, 51° 0′, 63° 46′, 72° 0′

Problem 3.22

Give an account of an X-ray diffraction method which has been used for the determination of Avogadro's number.

A monochromatic beam of X-rays is found to be reflected from a natural face of a lithium chloride crystal when the glancing angle (Bragg angle) is 17° 28′. The corresponding angle for a crystal of sodium chloride is found to be 15° 52′, and further investigation shows that two materials have the same crystal structure. What is the ratio of their molecular weights?

Density of lithium chloride = $2{\cdot}068 \text{ g cm}^{-3}$
Density of sodium chloride = $2{\cdot}163 \text{ g cm}^{-3}$

[Nat. Science Tripos, Pt. I, Cambridge, 1962]

Ans. 0·723

Problem 3.23

Describe experiments which establish (*a*) the particle nature of electromagnetic radiation, (*b*) the wave nature of electrons.

Electrons emerging from a hot filament with negligible velocities are accelerated through a potential of 1,000 volts and then diffracted by a crystal specimen. Determine the Bragg angle for first order diffraction from crystal planes of spacing 2·0 Å.

[B.Sc. (Technical), Manchester, 1963]

Ans. 5° 34′

Problem 3.24

Find the fraction of the volume occupied by hard spheres forming a face-centred cubic lattice.

What is the density of copper with a face-centred cubic lattice if its lattice constant is 3·61 Å? How many nearest neighbours does each copper atom have and how far are they from the reference atom?

At. wt. of copper = 63·54

[B.Sc. (General Hons.), Durham, 1961]

Ans. 8·97 g/cm^3; 12, 2·55 Å

Problem 3.25

Calculate the charge on an electron from the following measurements. A K_α-line, $\lambda = 0{\cdot}712 \times 10^{-10}$ m is reflected in first-order at

$\theta = 7° 15'$ from a NaCl crystal whose density is $2{\cdot}15 \times 10^3$ kg m^{-3}. The Faraday is $9{\cdot}6520 \times 10^7$ coulomb per kg atom. The atomic weight of Na is 23 and of Cl is 35·5.

[B.Sc. (Normal), Pt. I, Liverpool, 1963]

Ans. $1{\cdot}59 \times 10^{-19}$C

MISSING ORDERS

Problem 3.26

Give a history of one generation of neutrons, emitted from the fission of one ^{235}U nucleus in an atomic pile. Include a definition of the terms 'multiplication factor' and 'thermal neutron'.

A beam of thermal neutrons emitted from an opening in a pile, strikes the polished (001) face of an iron crystal at an angle of 43° 42′ and is strongly diffracted. Calculate the effective temperature of the pile. Would you expect any difference in intensity of the diffracted beam if the iron were magnetised?

(Mass of neutron $= 1{\cdot}675 \times 10^{-24}$ g. Lattice constant of α-iron $= 2{\cdot}861$ Å.)

[B.Sc. (Special), Sessional, Sheffield, 1953]

SOLUTION

From the crystal geometry the plane spacing is given by

$$d = \frac{a}{\sqrt{(h^2 + k^2 + l^2)}}$$

Since α-iron is a b.c.c. structure, then

$$n^2(h^2 + k^2 + l^2) = 2, 4, 6, 8, \text{etc.}$$

where n is the diffraction order. Thus from the (001) face we must get, at least, a second-order reflection i.e.

$$n^2(h^2 + k^2 + l^2) = 4$$

Now $$d_{(001)} = \frac{a}{\sqrt{1}} = 2{\cdot}861 \text{ Å}$$

so the effective wavelength λ_e of the neutrons is given by the Bragg equation

$$\lambda_e = \frac{2d \sin \theta}{n} = 2{\cdot}861 \sin 43^\circ\ 42' \text{ (since } n = 2)$$

$$= 1{\cdot}974\ \text{Å}$$

Also the effective wavelength of a neutron, by the de Broglie relation is $\lambda_e = \mathbf{h}/mv$ where v is its velocity, and from the kinetic theory we can equate its kinetic energy and its thermal energy. Thus

$$\tfrac{1}{2}mv^2 = \mathbf{k}T$$

where **k** is Boltzmann's constant and T is the effective temperature of the neutron. Therefore

$$\lambda_e = \frac{\mathbf{h}}{\sqrt{2m\mathbf{k}T}}$$

$$\text{or} \quad T = \frac{\mathbf{h}^2}{\lambda_e^2 2m\mathbf{k}}$$

$$= \frac{(6{\cdot}62 \times 10^{-27})^2}{(1{\cdot}974 \times 10^{-8})^2 \times 2 \times 1{\cdot}675 \times 10^{-24} \times 1{\cdot}38 \times 10^{-16}}$$

$$= \underline{243^\circ\text{K}}$$

Problem 3.27

A polycrystalline specimen of molybdenum (body-centred cubic with $a = 3{\cdot}14$ Å) is irradiated with a beam of X-rays (of wavelength 1·54 Å) in a cylindrical—film powder diffraction camera having a specimen—film distance of 10 cm. Calculate the position of the first diffraction maximum on the film.

Explain carefully how you would expect the diffraction maxima to change if (*a*) the average size of the crystallites were reduced, and (*b*) the specimen temperature were increased.

[B.Sc. (Ordinary), Manchester, 1963]

SOLUTION

The Bragg relation states

$$n\lambda = 2d \sin \theta$$

For the minimum value of θ we must have the lowest possible value of n/d. For a b.c.c. structure

$$d = \frac{a}{\sqrt{(h^2 + k^2 + 1^2)}}$$

where $n^2(h^2 + k^2 + l^2) = 2, 4, 6, 8$ etc. Therefore

$$\left(\frac{n}{d}\right)_{\min} = \frac{\sqrt{2}}{a}$$

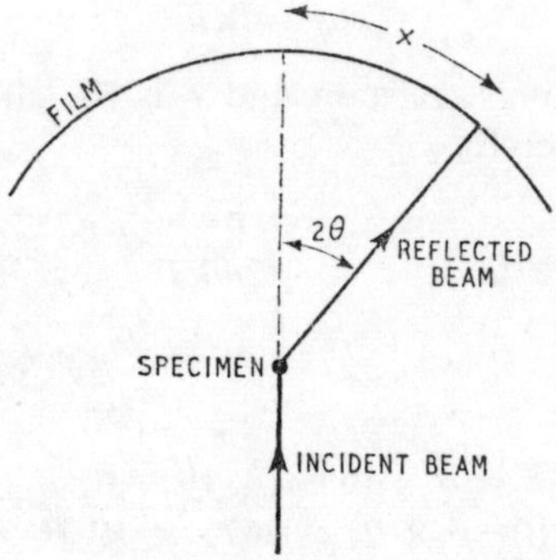

Fig. 3.2

The Bragg equation then gives

$$\sin\theta_{\min} = \frac{\sqrt{2}\lambda}{2a} = \frac{1{\cdot}54}{\sqrt{2} \times 3{\cdot}14}$$

or

$$\theta_{\min} = \underline{20^\circ\ 18' = 0{\cdot}3543 \text{ rad}}$$

The angle through which the beam is deviated is $2\theta_{\min}$ (Fig. 3.2). The distance of this maximum on the film from the straight through position is

$$x = 2\theta_{\min} r$$

where $\theta_{\min}$ is measured in radians and r is the radius of the film. Therefore

$$x = 2 \times 0{\cdot}3543 \times 10 = \underline{7{\cdot}09 \text{ cm}}$$

Problem 3.28

The lowest order d-values of α-iron, determined from an X-ray powder diffraction photograph, are:

1·99, 1·40, 1·14, 0·99, 0·89, 0·81, 0·75, 0·70 Å

Given that the crystal belongs to the cubic system, derive the crystal structure, and the size of the unit cell.

Determine also the number of atoms per unit cell. Hence derive the size of the unit cell from the following data and compare it with the result above.

Density of iron = $7{\cdot}85$ g cm^{-3}
Avogadro's number = $6{\cdot}02 \times 10^{23}$
At. wt. of iron = 55·8

[B.Sc. (Special), Sheffield, 1954]

SOLUTION

From the geometry of a cubic crystal, the crystal spacing

$$d = \frac{a}{\sqrt{(h^2 + k^2 + l^2)}}$$

The factor $(h^2 + k^2 + l^2)$ may, due to arithmetical possibilities, have only the values 1, 2, 3, 4, 5, 6, 8, 9 etc., but not 7, 15 or 23 etc. Thus an examination of the values of $1/d^2$ should enable $1/a^2$ to be determined.

The following table may be prepared:

d	$\frac{1}{d^2}$		$(h^2 + k^2 + l^2) \times \frac{1}{a_1^2}$		$(h^2 + k^2 + l^2) \times \frac{1}{a_2^2}$
1·99	0·252	=	1 × 0·252	=	2 × 0·126
1·40	0·510	=	2 × 0·255	=	4 × 0·128
1·14	0·770	=	3 × 0·257	=	6 × 0·129
0·99	1·02	=	4 × 0·255	=	8 × 0·128
0·89	1·26	=	5 × 0·252	=	10 × 0·126
0·81	1·525	=	6 × 0·254	=	12 × 0·127
0·75	1·78	=	7 × 0·254	=	14 × 0·127
0·70	2·04	=	8 × 0·255	=	16 × 0·128

The third column gives the values of $(h^2 + k^2 + l^2)$ for a value for $1/a^2 \simeq 0{\cdot}255$ but since this means that a value of 7 is necessary then the chosen value of $1/a^2$ must be wrong. The fourth column gives the values of $(h^2 + k^2 + l^2)$ for $1/a^2 \simeq 0{\cdot}1275$ and all these values are possible, and so we assume $1/a^2 = 0{\cdot}1275$, or $a = \underline{2{\cdot}80 \text{ Å}}$.

The values of $(h^2 + k^2 + l^2)$ which are present, viz., 2, 4, 6, 8 etc., indicate that the structure is body-centred.

In the b.c.c. structure, the eight atoms at the corners of the unit cube are each shared by eight unit cubes and so contribute $8 \times \frac{1}{8} = 1$

atom to each cube, the atom at the body centre belongs to that cube alone, and so there are in effect a total of 2 atoms per unit cube.

$$\text{The density of iron} = \frac{\text{mass of a unit cell}}{\text{volume of a unit cell}}$$

$$= \frac{2 \times 55{\cdot}8 \times \mathbf{m}_H}{a^3}$$

Therefore
$$a^3 = \frac{2 \times 55{\cdot}8}{7{\cdot}85 \times 6{\cdot}02 \times 10^{23}}$$

since $\mathbf{m}_H = 1/\mathbf{N}_0$, giving $a = \underline{2{\cdot}87\ \text{Å}}$.

Problem 3.29

Show that for an orthorhombic unit cell

$$\frac{1}{d^2} = \frac{h^2}{a^2} + \frac{k^2}{b^2} + \frac{l^2}{c^2}$$

where the symbols have their usual meaning.

Explain how X-ray diffraction by powder specimens may be used to distinguish between the three types of cubic Bravais lattices.

A powder photograph of a certain substance, taken with radiation of wavelength 1·54 Å, shows lines corresponding to the following Bragg angles

18°, 20·9°, 30·4°, 36·4°, 38·3°, 45·7°, 51·2°, 53·2°, 61·3°, 68·5°

Assuming the substance is cubic, determine the unit cell translation and show that the lattice must be face-centred.

[B.Sc. (Technical), Manchester, 1963]

SOLUTION

From the geometry of a cubic crystal ($a = b = c$)

$$\frac{1}{d^2} = \frac{h^2 + k^2 + l^2}{a^2}$$

The factor $(h^2 + k^2 + l^2)$ may have values 1, 2, 3, 4, 5, etc., but not 7, 15 or 23 etc. As indicated in the previous example an examination of the values $1/d^2$ for a crystal therefore enables $1/a^2$ to be determined.

The Bragg relation states

$$n\lambda = 2d \sin \theta$$

$$\therefore \qquad \sin^2 \theta = \frac{n^2\lambda^2}{4d^2}$$

or

$$\sin^2 \theta = \frac{\lambda^2}{4a^2} n^2(h^2 + k^2 + l^2)$$

Consider the following table:

θ	18°	20·9°	30·4°	36·4°	38·3°	45·7°	51·2°	53·2°	61·3°	68·5°
$\sin \theta$	0·309	0·357	0·506	0·593	0·620	0·716	0·779	0·801	0·877	0·930
$\sin^2 \theta$	0·0955	0·127	0·256	0·352	0·384	0·512	0·607	0·641	0·796	0·866
$\sin^2 \theta$	3 × 0·0318	4 × 0·0318	8 × 0·032	11 × 0·032	12 × 0·032	16 × 0·032	19 × 0·032	20 × 0·032	24 × 0·0318	27 × 0·0318

This indicates that $n^2(h^2 + k^2 + l^2)$ assumes the values 3, 4, 8, 11, 12, 16 etc. The structure must therefore be face-centred cubic. Also the table indicates that

$$\frac{\lambda^2}{4a^2} = 0{\cdot}032$$

giving

$$a = \frac{\lambda}{2} \frac{1}{\sqrt{0{\cdot}032}} = \frac{1{\cdot}54}{2 \times 0{\cdot}179} = \underline{4{\cdot}3\ \text{Å}}$$

Problem 3.30

Monochromatic X-rays of wavelength 1·54 Å are incident on a powder specimen of an element having a body-centred structure with $a = 2{\cdot}94$ Å. Calculate the angles and relative intensities of the first two rings of the resultant Debye-Scherrer pattern, ignoring the angular variation of the atomic scattering factor.

The alloy CuZn has a body-centred structure as above and can exist in either the disordered state (in which the atoms are randomly arranged) or in the ordered state in which a Cu atom is at the body centre of a cube of Zn atom). Describe and explain qualitatively the difference that will be observed in the powder patterns of the alloy in the two states. The atomic numbers of Cu and Zn are 30 and 31 respectively.

What difference would be observed in similar patterns obtained with monochromatic neutrons of the same wavelength as the X-rays? Neutron scattering cross sections are 7 barns for Cu and 4 barns for Zn.

[B.Sc. (Honours), Pt. II, Manchester, 1963]

Ans. 43° 30′, 63° 12′
Ratio of intensities = $\sqrt{2}/1$

MISCELLANEOUS

Problem 3.31

Explain under what conditions it is possible to reflect X-rays from polished metal surfaces and show how this phenomenon is utilised for the absolute determination of X-ray wavelengths.

A polished metal grating has 10,000 lines/cm and is illuminated at grazing incidence with a narrow beam of X-rays of wavelength 1·54 Å. Calculate the dispersion of the grating for the first order diffraction maximum, and comment on your result.

[B.Sc. (Special), Sessional, Sheffield, 1963]

SOLUTION

With *N*, *P*, *R*, *S* representing the ruled grating the X-ray beam, incident from the left, is diffracted as shown in Fig. 3.3.

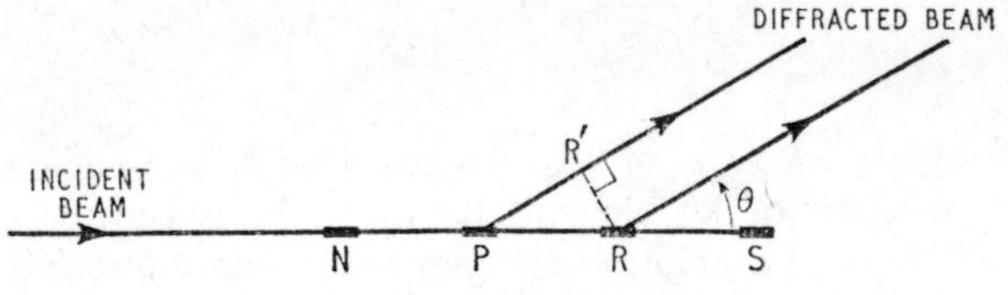

Fig. 3.3

In the first order the path difference $PR - PR' = \lambda$. If the grating spacing is a, we have

$$a - a\cos\theta = \lambda$$

or
$$1 - \cos\theta = \frac{\lambda}{a} = \frac{1{\cdot}54 \times 10^{-8}}{10^{-4}} = 1{\cdot}54 \times 10^{-4}$$

θ is therefore very small and we may write $\cos\theta = 1 - \theta^2/2$. Thus

$$\frac{\theta^2}{2} = \frac{\lambda}{a} \quad \text{or} \quad \theta = \left(\frac{2\lambda}{a}\right)^{\frac{1}{2}}$$

The dispersion

$$\frac{d\theta}{d\lambda} = \left(\frac{1}{2\lambda a}\right)^{\frac{1}{2}}$$

$$= \frac{1}{(2 \times 1{\cdot}54 \times 10^{-8} \times 10^{-4})^{\frac{1}{2}}} = 5{\cdot}7 \times 10^5 \text{ rad/cm}$$

$$= \underline{5{\cdot}7 \times 10^{-3} \text{ rad/(Å)}}$$

Problem 3.32

Describe and explain the X-ray powder method for the identification of substances and discuss its advantages and limitations.

Explain the effect of the size of the crystals on the width of the lines on a powder photograph and calculate the broadening due to this effect alone, of a line at a Bragg angle of 40° for crystals whose size is 250 Å.

Assume that the wavelength of the X-rays is 1·54 Å.

Indicate the difficulties of using this idea to determine the crystal size of a sample of material.

[Dip. Tech. Pt. I, Salford, 1962]

SOLUTION

For reflection $n\lambda = 2d \sin\theta$. Let this condition be satisfied along the direction CBA in Fig. 3.4. If the crystallite is $2\,m$ planes thick and the point B is on the middle plane, then the total path difference between rays from B and A is $mn\lambda$.

If the angle $d\theta$ is such that the path difference between rays from B' and A is $(mn + \frac{1}{2})\lambda$, then the rays reflected from the plane $2m$ will be completely out of phase with that from plane m, and will destructively interfere. Similarly the rays from planes $2m - 1$, $2m - 2$, etc. will destructively interfere with the rays from planes $m - 1$, $m - 2$, etc. Thus ray $C'B'A$ represents the limit of the X-ray line. Now

$$mn\lambda = 2md \sin\theta \qquad (3.32.1)$$

and

$$(mn + \tfrac{1}{2})\lambda = 2md \sin(\theta + d\theta) \qquad (3.32.2)$$

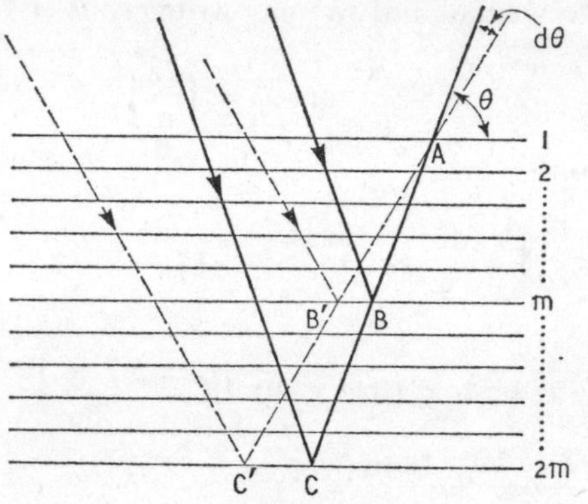

Fig. 3.4

Subtracting Equation 3.32.1 from 3.32.2

$$\begin{aligned}\lambda &= 2md\{\sin(\theta + d\theta) - \sin\theta\}\\ &= 2md(\sin\theta\cos d\theta - \cos\theta\sin d\theta - \sin\theta)\\ &= 2md\cos\theta\, d\theta\end{aligned}$$

$$\therefore \qquad d\theta = \frac{\lambda}{4md\cos\theta} = \frac{\lambda}{2t\cos\theta}$$

where t is the thickness of the crystallite. Therefore

$$d\theta = \frac{1{\cdot}54}{2 \times 250\cos 40^\circ} = 4{\cdot}02 \times 10^{-3}\text{ rad}$$

The total width of the X-ray line due to crystal size is therefore $8{\cdot}04 \times 10^{-3}$ rad.

4
Solid State Physics

LATTICE ENERGY OF IONIC CRYSTALS

Problem 4.1

Give a brief account of the influence of inter-atomic forces on the stereochemistry of solid elements and single compounds.

Body-centred CsCl has a cubic unit cell of side of 4·11 Å and an energy of cohesion of 6·30 eV per ion pair. If the potential energy of repulsion between a pair of isolated ions at a separation r is given by $A \exp(-r/\rho)$, calculate the constants A and ρ, assuming that in the solid state the energy of repulsion is only significant between nearest neighbours.

What experimental information is required to obtain an estimate of the energy of cohesion?

(Madelung's constant is 1·763 for this structure, referred to anion–cation separation.)

[B.Sc. (Special), Pt. II, London, 1961]

SOLUTION

For two isolated ions the potential energy of repulsion

$$\varepsilon_{\text{rep}} = A \exp\left(-\frac{r}{\rho}\right)$$

Each ion of CsCl (b.c.c.) has 8 nearest neighbours so the total energy of repulsion equals $8A \exp(-r/\rho)$ per ion.

The potential energy ε_c due to Coulomb attraction is given by

$$\varepsilon_c = -M\frac{e^2}{r}$$

per ion, where M is the Madelung constant. Thus the total energy $E(r)$ of an ion due to the presence of the other ions is given by

$$E(r) = -\frac{Me^2}{r} + 8A \exp\left(-\frac{r}{\rho}\right) \qquad (4.1.1)$$

The total binding energy (or energy of cohesion) of the crystal containing N positive ions and N negative ions is then

$$NE(r) = N\left[-\frac{Me^2}{r_0} + 8A \exp\left(-\frac{r_0}{\rho}\right)\right]$$

where r_0 is the equilibrium separation of the ions. The energy of cohesion per ion pair

$$-E(r) = 6{\cdot}30 \text{ eV} = \frac{Me^2}{r_0} + 8A \exp\left(-\frac{r_0}{\rho}\right) \qquad (4.1.2)$$

If the side of the unit cell is a then the separation of nearest neighbours

$$r_0 = \frac{a\sqrt{3}}{2} = 3{\cdot}56 \text{ Å}$$

Substitution into Equation 4.1.2. gives

$$-6{\cdot}30 \times 1{\cdot}6 \times 10^{-12} = -\frac{1{\cdot}763 \times (4{\cdot}8 \times 10^{-10})^2}{3{\cdot}56 \times 10^{-8}} + 8A \exp\left(\frac{-3.56 \times 10^{-8}}{\rho}\right)$$

$$\therefore \qquad A \exp\left(\frac{-3{\cdot}56 \times 10^{-8}}{\rho}\right) = 4 \times 10^{-14} \qquad (4.1.3)$$

The equilibrium separation of the ions is for $dE(r)/dr = 0$. Thus, differentiating Equation 4.1.1. and putting $r = r_0$ we have

$$\frac{Me^2}{r_0^2} - \frac{8A}{\rho} \exp\left(\frac{-r_0}{\rho}\right) = 0$$

Numerical substitution gives

$$\frac{A}{\rho} \exp\left(-\frac{3{\cdot}56 \times 10^{-8}}{\rho}\right) = 4{\cdot}006 \times 10^{-5} \qquad (4.1.4)$$

Dividing Equation 4.1.3 by 4.1.4,

$$\rho \simeq \frac{4 \times 10^{-14}}{4{\cdot}006 \times 10^{-5}} = 1{\cdot}00 \times 10^{-9} \text{ cm}$$

Substituting this value of ρ into Equation 4.1.3 and re-arranging we have

$$A \simeq 4 \times 10^{-14} \exp\left(\frac{3{\cdot}56 \times 10^{-8}}{1{\cdot}00 \times 10^{-9}}\right) \simeq \underline{100}$$

Problem 4.2

Define the term 'cohesive energy' of a solid and show how it may be estimated theoretically for an ionic salt of known structure.

LiF has a face-centred cubic unit cell of side 4·02 Å at 0°K. Madelung's constant for this structure is 1·747 referred to the anion–cation separation r. If the repulsive potential between adjacent ions is proportional to r^{-n} estimate a value for n given that the compressibility is $-1{\cdot}43 \times 10^{-12}$ cm^2 dyne^{-1} at 0°K.

[B.Sc. (Special), Pt. II, London, 1962]

SOLUTION

Compressibility K is defined as

$$K = -\frac{1}{V}\frac{dV}{dP} \tag{4.2.1}$$

In the region of absolute zero, specific heat has a vanishingly small value so that if a substance is compressed its gain in internal energy $dE = -P\,dV$ which gives

$$\frac{d^2E}{dV^2} = -\frac{dP}{dV}$$

thus (from Equation 4.2.1) at 0°K the compressibility K_0 is given by

$$\frac{1}{K_0} = V\frac{d^2E}{dV^2} \tag{4.2.2}$$

Now
$$\frac{dE}{dV} = \frac{dE}{dr}\cdot\frac{dr}{dV}$$

giving
$$\frac{d^2E}{dV^2} = \frac{dE}{dr}\frac{d^2r}{dV^2} + \frac{d^2E}{dr^2}\left(\frac{dr}{dV}\right)^2$$
$$= \frac{d^2E}{dr^2}\left(\frac{dr}{dV}\right)^2 \tag{4.2.3}$$

since $dE/dr = 0$ at the equilibrium lattice spacing.

We must now determine the values of the two factors in Equation 4.2.3. For the f.c.c. structure there is, in effect, half a molecule of LiF per volume r^3, and so N molecules occupy a volume

$$V = 2Nr^3 \tag{4.2.4}$$

Thus

$$\left(\frac{dV}{dr}\right)^2 = (6Nr^2)^2 \tag{4.2.5}$$

As seen in Problem 4.1 the total internal energy E of the solid is given by the sum of the coulomb and the repulsive energies, i.e.

$$E = N\left(-\frac{Me^2}{r} + Br^{-n}\right) \tag{4.2.6}$$

where M is the Madelung constant and B a constant of proportionality. Differentiating Equation 4.2.6 we get

$$\frac{dE}{dr} = N\left\{\frac{Me^2}{r^2} - nBr^{-(n+1)}\right\} \tag{4.2.7}$$

and

$$\frac{d^2E}{dr^2} = N\left\{-\frac{2Me^2}{r^3} + n(n+1)Br^{-(n+2)}\right\} \tag{4.2.8}$$

Remembering that $dE/dr = 0$ at the equilibrium separation, Equation 4.2.7 gives

$$B = \frac{Me^2}{nr^{-(n-1)}} \tag{4.2.9}$$

Substituting this value of B into Equation 4.2.8 yields

$$\frac{d^2E}{dr^2} = \frac{NMe^2}{r^3}(n-1) \tag{4.2.10}$$

We have now obtained expressions for both factors in Equation 4.2.3 so that from Equations 4.2.2 and 4.2.3

$$\frac{1}{K_0} = V\frac{d^2E}{dV^2} = V\frac{d^2E}{dr^2}\left(\frac{dr}{dV}\right)^2$$

Thus, from Equations 4.2.4, 4.2.5 and 4.2.10

$$\frac{1}{K_0} = 2Nr^3 \cdot \frac{NMe^2}{r^3}(n-1)\frac{1}{(6Nr^2)^2}$$

$$= \frac{Me^2(n-1)}{18r^4}$$

On re-arranging

$$n = \frac{18r^4}{K_0 M e^2} + 1$$

Since the anion–cation separation $r = \frac{1}{2} \times 4{\cdot}02$ Å, numerical substitution gives

$$n = \frac{18 \times (2{\cdot}01 \times 10^{-8})^4}{1{\cdot}43 \times 10^{-12} \times 1{\cdot}747 \times (4{\cdot}8 \times 10^{-10})^2} + 1 = \underline{6{\cdot}1}$$

MAGNETIC PROPERTIES OF MATERIALS

Problem 4.3

An aqueous solution containing 25 g of $NiSO_4$ in 100 g of solution is contained in a suitable **U**-tube. It is found that at a temperature of 18°C the maximum height of the column of solution which can be supported by magnetic pressure in a uniform field of 20,000 gauss is 1·3 cm. Under similar conditions the surface of pure water is depressed in the same field by 1·46 mm. Neglecting the paramagnetic effect of air and the diamagnetic effect of the sulphate ion, calculate the magnetic moment of the Ni^{++} ion in terms of the Bohr magneton.

[B.Sc. (Honours), Liverpool, 1960]

SOLUTION

The mass susceptibilities of water and of the solution may be found by considering the changes in the liquid levels when the field is applied. The theory of the Quincke method shows that the mass susceptibility χ of the liquid in the **U**-tube is given by

$$\chi = \frac{2gh}{H^2}$$

where H is the magnetic field which produces a difference in surface levels of h.

For the solution

$$\chi_{sol} = \frac{2 \times 980 \times 1{\cdot}3}{(2 \times 10^4)^2} = 6{\cdot}37 \times 10^{-6} \text{ e.m.u. g}^{-1}$$

and for the water

$$\chi_{\text{water}} = -\frac{2 \times 980 \times 0{\cdot}146}{(2 \times 10^4)^2} = -0{\cdot}715 \times 10^{-6} \text{ e.m.u. g}^{-1}$$

The mass susceptibility of the solution may be expressed in terms of the mass susceptibilities of nickel sulphate and of water. Thus

$$(M + m)\chi_{\text{sol}} = m\chi_{\text{NiSO}_4} + M\chi_{\text{water}}$$

where the solution contains m gram of $NiSO_4$ and M gram of water. In this case $m = 25$ g and $(M + m) = 100$ g giving $M = 75$ g. Thus

$$\chi_{\text{NiSO}_4} = \frac{M + m}{m}\chi_{\text{sol}} - \frac{M}{m}\chi_{\text{water}}$$

and substitution gives

$$\chi_{\text{NiSO}_4} = \frac{100}{25} \times 6{\cdot}37 \times 10^{-6} + \frac{75}{25} \times 0{\cdot}715 \times 10^{-6}$$

$$= 27{\cdot}63 \times 10^{-6} \text{ e.m.u. g}^{-1}$$

The susceptibility of the nickel sulphate may be expressed in terms of the susceptibilities of the constituent Ni^{++} and SO_4^{--} ions. The molecular weight of $NiSO_4$ is 154·7 and the ionic weights of Ni^{++} and SO_4^{--} are 58·7 and 96 respectively, so in terms of mass susceptibilities

$$154{\cdot}7\ \chi_{\text{NiSO}_4} = 58{\cdot}7\ \chi_{\text{Ni}^{++}} + 96\ \chi_{\text{SO}_4^{--}}$$

But since $\chi_{\text{SO}_4^{--}}$ is negligible we have

$$\chi_{\text{Ni}^{++}} = \frac{154{\cdot}7}{58{\cdot}7} \times \chi_{\text{NiSO}_4}$$

$$= \frac{154{\cdot}7 \times 27{\cdot}63 \times 10^{-6}}{58{\cdot}7} = 72{\cdot}8 \times 10^{-6} \text{ e.m.u. g}^{-1}$$

From the Langevin theory of paramagnetism the mass susceptibility of a substance is given by

$$\chi = \frac{N\mu^2}{3\mathbf{k}T}$$

where N is the number of atomic dipoles per gram, μ is the magnetic moment of each dipole, **k** is Boltzmann's constant and T is the absolute temperature. Therefore

$$\mu^2 = \frac{3\mathbf{k}T\chi}{N} = \frac{3 \times 1{\cdot}38 \times 10^{-16} \times 291 \times 72{\cdot}8 \times 10^{-6}}{\dfrac{6{\cdot}025 \times 10^{23}}{58{\cdot}7}} \text{ (e.m.u.)}^2$$

$$\therefore \qquad \mu = 2{\cdot}92 \times 10^{-20} \text{ e.m.u.}$$

$$= \frac{2{\cdot}92 \times 10^{-20}}{0{\cdot}927 \times 10^{-20}} = \underline{3{\cdot}14\,\mu_B}$$

Problem 4.4

What is the Bohr magneton? Discuss briefly the origin of atomic moments.

The saturation intensity of magnetisation of nickel is 517 c.g.s. units per unit volume. Assuming that nickel has a face-centred cubic structure with a cube edge of 3·5 Å, calculate the effective number of Bohr magnetons per atom.

[Dip. Tech. Elec. Eng., Salford, 1963]

SOLUTION

In a face-centred cubic structure there are, in effect, 4 atoms per unit cube. The volume of one unit cube $= (3{\cdot}5 \times 10^{-8})^3$ cm³ so that

$$\text{number of Ni atoms per unit volume} = \frac{4}{(3{\cdot}5 \times 10^{-8})^3} \text{ cm}^{-3}$$

$$= 9{\cdot}33 \times 10^{22} \text{ cm}^{-3}$$

Thus

$$\text{magnetic moment per atom} = \frac{517}{9{\cdot}33 \times 10^{22} \times 0{\cdot}927 \times 10^{-20}}\,\mu_B$$

$$= \underline{0{\cdot}60\,\mu_B}$$

Problem 4.5

Describe a method suitable for the determination of the saturation intensity of magnetisation of a ferromagnetic material over a wide range of temperature.

Explain the effect on the saturation intensity of nickel produced by diluting it with copper.

The saturation intensity of nickel at 20°C is 54·3 e.m.u. g^{-1}. Calculate the saturation intensity of the alloy when (*a*) 5 atomic per cent of copper and (*b*) 5 atomic per cent of zinc, are added, given that each nickel atom has an effective Bohr magneton number of 0·6.

[Dip. Tech. Pt. II, Salford, 1963]

SOLUTION

(*a*) Each Cu atom has 11 electrons to be distributed between the 3*d* and 4*s* bands, while each nickel atom only provides 10 electrons for these bands. When a nickel atom is replaced by a copper atom the extra electron takes up a place in the 'magnetic' 3*d* band, which being more than half filled reduces the magnetic moment of the solid by $1\,\mu_B$.

For pure nickel, it is given that 100 atoms have a moment of $60\,\mu_B$ and so if we have 5% of copper atoms present, these 5 copper atoms reduce this moment by $5\mu_B$.

The 100 atoms of alloy have a moment of only $55\,\mu_B$. Thus

$$\text{saturation intensity of } 5\% \text{ Cu alloy} = 54{\cdot}3 \times \frac{55}{60}$$

$$= \underline{49{\cdot}7 \text{ e.m.u. g}^{-1}}$$

(*b*) Similarly each atom of zinc contributes 2 electrons to the 3*d* band and so reduces the magnetic moment by $2\mu_B$. Thus

$$\text{saturation intensity of Zn alloy} = 54{\cdot}3 \times \frac{50}{60}$$

$$= \underline{45{\cdot}2 \text{ e.m.u. g}^{-1}}$$

Problem 4.6

Account briefly for the magnetic properties of ferrites. In a sample of the ferrite $M^{++}Fe_2^{+++}O_4$, one sixth of the M^{++} ions are Zn^{++} ions, and five-sixths are Fe^{++} ions. Estimate the effective number of Bohr magnetons per molecule of $M^{++}Fe^{+++}\,O_4$.

Fe^{++} ions have 6 electrons in the 3*d* band;

Fe^{+++} ions have 5 electrons in the 3*d* band;
Zn^{++} ions have 10 electrons in the 3*d* band.

[H.N.C., Appled Physics, Salford, 1963]

SOLUTION

The 3*d* band can accommodate a maximum of 10 electrons per atom, thus for Fe^{++} ions there are four 3*d* electrons with unpaired spins giving an atomic moment of $4\mu_B$ per ion and for Fe^{+++} ions which have five electrons with unpaired spins in the 3*d* band the magnetic

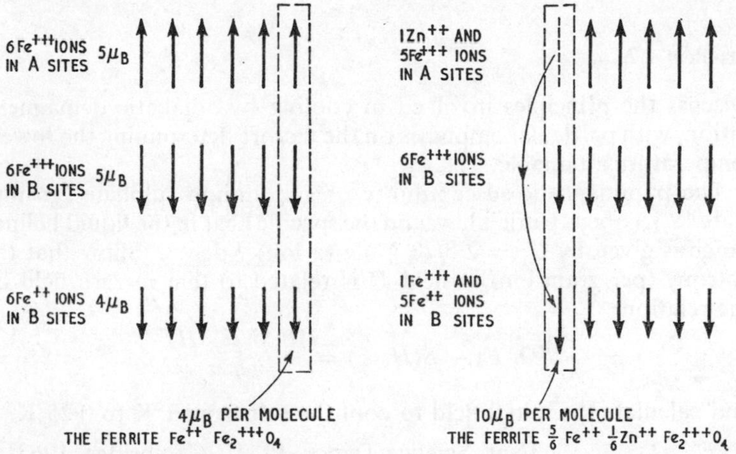

Fig. 4.1

moment is $5\mu_B$ per ion. Since in the case of Zn^{++} the 3*d* band is completely filled, the Zn^{++} ions have no permanent magnetic moments.

The spinel crystal structure with 8 tetrahedral '*A*' sites and sixteen octahedral '*B*' sites per unit cell, has strong antiferromagnetic interaction between ions at these two types of lattice site. In the ferrite $Fe^{++}Fe_2^{+++}O_4$ half of the Fe^{+++} ions are at *A* sites and half at *B* sites, and as these are in antiferromagnetic array they provide no resultant moment, and so the resultant moment for this ferrite is due to the Fe^{++} ions alone i.e. $4\mu_B$ per molecule (see Fig. 4.1).

The Zn^{++} ion, while replacing a Fe^{++} ion, prefers to go to an *A* site, and the Fe^{+++} ion it displaces is forced to the *B* site previously occupied by the Fe^{++} ion. This displaced Fe^{+++} ion is now set antiparallel to the *A* site Fe^{+++} ions and parallel to the ions in the *B* sites. Thus considering the molecule containing the Zn^{++} ion we now have two parallel Fe^{+++} ions, i.e. a magnetic moment of $10\,\mu_B$ per molecule (see Fig. 4.1).

This situation will hold providing the concentration of Zn^{++} ions is not too high, and it should be valid in the present example.

Thus for the ferrite $\frac{1}{6}Zn^{++}\,\frac{5}{6}Fe^{++}Fe_2O_4$, the effective number of Bohr magnetons per molecule is

$$\frac{1}{6} \times 10 + \frac{5}{6} \times 4 = \underline{5\,\mu_B \text{ per molecule}}$$

Problem 4.7

Discuss the principles involved in cooling by adiabatic demagnetisation, with particular emphasis on the factors determining the lowest temperature attainable.

The paramagnetic susceptibility of gadolinium sulphate (ground state $^8S_{7/2}$) obeys Curie's law, and the specific heat in the liquid helium range is given by $C_p = 2{\cdot}5/T^2$ joule (g ion)$^{-1}$ deg^{-1}. Show that the entropy (per gram ion) in field H is related to that in zero field by the relation

$$S(O, T) - S(H, T) = \frac{10{\cdot}5N\mu_B{}^2H^2}{kT^2}$$

and calculate the initial field to cool the salt from 1°K to 0·25°K.

[Nat. Science Tripos, Pt. II, Cambridge, 1962]

SOLUTION

From the first law of thermo-dynamics

$$\mathrm{d}Q = \mathrm{d}U + \mathrm{d}W$$

or in this case

$$T\,\mathrm{d}S = C_H\,\mathrm{d}T - M\,\mathrm{d}H \qquad (4.7.1.$$

where $\mathrm{d}S$ is the increase in entropy due to an increase in temperature $\mathrm{d}T$ and an increase in magnetic field $\mathrm{d}H$, C_H being the specific heat at constant magnetic field, and M being the magnetic moment per unit mass. At constant temperature we have

$$\mathrm{d}S = -\frac{M\,\mathrm{d}H}{T}$$

and integration gives

$$\int_{S(O,T)}^{S(H,T)} \mathrm{d}S = -\int_0^H \frac{M}{T}\,\mathrm{d}H$$

or

$$S(O, T) - S(H, T) = \int_0^H \frac{M}{T}\,\mathrm{d}H$$

But $M = \chi H$ where χ is the mass susceptibility and by the Langevin theory

$$\chi = \frac{N\mu^2}{3\mathbf{k}T}$$

$$\therefore \qquad S(O, T) - S(H, T) = \int_0^H \frac{N\mu^2}{3\mathbf{k}T^2} H\,\mathrm{d}H = \frac{N\mu^2H^2}{6\mathbf{k}T^2}$$

The atomic magnetic moment $\mu = g\mu_B\sqrt{J(J+1)}$ where g is the Landé splitting factor. For the ground state of gadolinium sulphate $J = 7/2$, $L = 0$ and $S = 7/2$, and using

$$g = 1 + \frac{J(J+1) + S(S+1) - L(L+1)}{2J(J+1)}$$

we have

$$\mu = \sqrt{63}\,\mu_B$$

Substitution gives

$$S(O, T) - S(H, T) = \frac{10{\cdot}5\, N\mu_B^2H^2}{\mathbf{k}T^2}$$

Taking the entropy of the system to be a function of H and T we can write

$$\mathrm{d}S = \left(\frac{\partial S}{\partial T}\right)_H \mathrm{d}T + \left(\frac{\partial S}{\partial H}\right)_T \mathrm{d}H \qquad (4.7.2)$$

and for the adiabatic process $\mathrm{d}S = 0$. From Equation 4.7.1 above

$$T\left(\frac{\partial S}{\partial T}\right)_H = C_H \quad \text{and} \quad T\left(\frac{\partial S}{\partial H}\right)_T = -M$$

Thus Equation 4.7.2 becomes

$$C_H\,\mathrm{d}T - M\,\mathrm{d}H = 0$$

Substituting for C_H and M we have, since $C_H = C_p$

$$\frac{2{\cdot}5 \times 10^7\, \mathrm{d}T}{T^2} - \frac{N\mu^2 H}{3\mathbf{k}T}\,\mathrm{d}H = 0$$

or
$$2{\cdot}5 \times 10^7 \frac{\mathrm{d}T}{T} = \frac{N\mu^2}{3\mathbf{k}} H\,\mathrm{d}H$$

Integrating between the appropriate limits we have

$$2{\cdot}5 \times 10^7 \int_{1\cdot0}^{0\cdot25} \frac{\mathrm{d}T}{T} = \int_{H}^{0} \frac{N\mu^2}{3\mathbf{k}} H\,\mathrm{d}H$$

giving
$$2{\cdot}5 \times 10^7 \log_e \tfrac{1}{4} = -\frac{N\mu^2}{6\mathbf{k}} H^2$$

or
$$H^2 = \frac{2{\cdot}5 \times 10^7 \log_e 4 \times 6 \times 1{\cdot}38 \times 10^{-16}}{6{\cdot}025 \times 10^{23} \times 63 \times (0{\cdot}927 \times 10^{-20})^2}$$

i.e.
$$H = \underline{2{,}970 \text{ oersted}}$$

Problem 4.8

The Langevin theory of a paramagnetic gas shows that the molecular susceptibility $\chi_M = \sigma_0^2/3RT$ where σ_0 is the saturation intensity of magnetisation per gram-molecule at absolute zero, R is the gas constant and T the absolute temperature. Indicate briefly the physical ideas leading to this expression.

Show how this has to be modified to account for the observed magnetic susceptibility of the paramagnetic salts of the iron and rare earth groups.

The molecular susceptibility of anhydrous manganous sulphate at 17°C is found to be $14{\cdot}32 \times 10^{-3}$ e.m.u. If the spectroscopic state of the Mn^{++} ion is $^6S_{\frac{5}{2}}$ and the gram-ionic susceptibility of the SO_4^{--} ion is $-33{\cdot}6 \times 10^{-6}$ e.m.u., calculate the magnetic moment of the manganese ion in Bohr magnetons and compare it with its theoretical value.

[Dip. Tech. Pt. II, Salford, 1962]

Ans. $5{\cdot}77\,\mu_B$, $5{\cdot}92\,\mu_B$ (theoretical)

POINT DEFECTS

Problem 4.9

Explain what is meant by Schottky and by Frenkel defects in a crystal lattice and indicate briefly how the physical properties of an ionic crystal are modified by the presence of such defects.

Estimate the number of Schottky defects in a 1 cm cube of rocksalt at room temperature, assuming that the distance between adjacent ions in sodium chloride is 2·81 Å and that the energy to remove a pair of oppositely-charged ions from the lattice is 2 eV.

(Stirlings' formula states that $\log_e (N!) = N(\log_e N - 1)$ for N large.)

[B.Sc. (Ordinary), Manchester, 1963]

SOLUTION

If there are N positive and N negative ions and n pairs of vacancies, then the configurational entropy of the positive ions is $\mathbf{k} \log P$, where P is the number of ways of arranging the n vacant ionic sites amongst the $(N + n)$ available sites; the same applies to the negative ions. Thus the total configurational entropy

$$S = 2\mathbf{k} \log_e P$$

where
$$P = \frac{(N+n)!}{n!N!}$$

i.e.
$$S = 2\mathbf{k} \log_e \left\{\frac{(N+n)!}{n!N!}\right\}$$

The free energy $F = U - TS$ where U is the energy to create the n vacancy pairs. Thus

$$F = n\phi - 2\mathbf{k}T \log_e \left\{\frac{(N+n)!}{N!n!}\right\}$$

where ϕ is the energy to create one vacancy pair (i.e. $\phi = 2$ eV) and applying Stirling's formula we get

$$F = n\phi - 2\mathbf{k}T\{(N+n)\log(N+n) - N\log_e N - n\log_e n\}$$

At equilibrium the free energy is a minimum i.e. $\mathrm{d}F/\mathrm{d}n = 0$, therefore

$$\phi - 2\mathbf{k}T\{\log(N+n) - \log n\} = 0$$

or
$$\frac{n}{N+n} = \exp\left(-\frac{\phi}{2\mathbf{k}T}\right)$$

or since $n \ll N$

$$\frac{n}{N} = \exp\left(-\frac{\phi}{2kT}\right)$$

The distance between adjacent ions (Na^+Cl^-) is 2·81 Å thus, the cube side of the unit face-centred cell of rock salt,

$$a = 2 \times 2{\cdot}81 = 5{\cdot}62 \text{ Å}$$

Each such cell contains, in effect, 4 positive and 4 negative ions. The number of positive ions per unit volume N is given by

$$N = \frac{4}{a^3} = \frac{4}{(5{\cdot}62 \times 10^{-8})^3} = \underline{2{\cdot}25 \times 10^{22} \text{ cm}^{-3}}$$

Assuming that room temperature is 295°K then

$$\frac{n}{N} = \exp(-2 \times 1{\cdot}6 \times 10^{-12}/1{\cdot}38 \times 10^{-16} \times 295 \times 2)$$

$$= \exp(-39{\cdot}3) \simeq 9{\cdot}0 \times 10^{-18}$$

Therefore the number of vacancy pairs $n \simeq \underline{2 \times 10^5 \text{ cm}^{-3}}$.

Problem 4.10

Discuss the relation between entropy and probability.

Explain carefully how the presence of vacancies results in an increase of entropy of a crystal, deduce an expression for the concentration of vacancies in thermal equilibrium at any temperature, and calculate this concentration for aluminium at 600°C. Discuss the experimental evidence for the existence of vacancies in metals.

(Energy of formation of vacancy in aluminium = 0·8 eV; entropy of formation $\simeq 2\,k$; $k = 8{\cdot}6 \times 10^{-5}$ eV/deg.)

[Nat. Science Tripos, Pt. II, Cambridge, 1962]

Ans. $1{\cdot}8 \times 10^{-4}$

THE FREE ELECTRON AND ZONE THEORIES

Problem 4.11

By considering the distribution of free electrons in k-space for a cubical specimen of metal show that the Fermi energy at 0°K is given by

$$E_F = \frac{\hbar^2}{2m}(3\pi^2 n)^{2/3}$$

where n is the concentration of free electrons.

Calculate E_F for Na in which the nearest neighbour distance is 3·71 Å. Discuss the relationship between the Fermi sphere and the Brillouin zone boundary with particular reference to conduction phenomena in monovalent and divalent metals.

[B.Sc. (Honours), Pt. II, Durham, 1962]

SOLUTION

Sodium has a body-centred cubic structure, and so has, in effect, two atoms to each unit cell. Also if the nearest neighbour distance is d, then the side of the unit cell

$$a = \frac{2d}{\sqrt{3}} = \frac{2 \times 3{\cdot}71 \times 10^{-8}}{\sqrt{3}}$$

The density of atoms in the metal is thus

$$\frac{2}{a^3} = \frac{2 \times 3\sqrt{3}}{8 \times (3{\cdot}71 \times 10^{-8})^3} = 2{\cdot}54 \times 10^{22}\ \text{cm}^{-3}$$

Since sodium is monovalent, then the density of free electrons n is also $2{\cdot}54 \times 10^{22}\ \text{cm}^{-3}$.

The Fermi level E_F is given by

$$E_F = \frac{\hbar^2}{2m}(3\pi^2 n)^{\frac{2}{3}}$$

Substituting numerical data we have

$$E_F = \frac{(6{\cdot}62 \times 10^{-27})^2}{8\pi^2 \times 9{\cdot}11 \times 10^{-28}} \times (3\pi^2 \times 2{\cdot}54 \times 10^{22})^{\frac{2}{3}}$$

$$= 5{\cdot}04 \times 10^{-12}\ \text{erg}$$

$$= \frac{5{\cdot}04 \times 10^{-12}}{1{\cdot}60 \times 10^{-12}} = \underline{3{\cdot}15\ \text{eV}}$$

Problem 4.12

Discuss briefly the wave functions and energy levels of an electron in a periodic potential, explaining how the states are grouped into bands, each band, in a finite specimen of lattice, containing twice as many states as there are unit cells.

A certain crystal has unit cells which are cubes of side 3×10^{-8} cm. If the periodic part of the potential acting on a conduction electron could be considered very weak, what would be the kinetic energy of a state at the top of the lowest band (in electron-volts)?

[B.Sc. (Honours), Liverpool, 1963]

SOLUTION

For a simple cubic structure, the first zone in k-space is a cube of half side π/a (where a is the lattice spacing). If we have very weak binding, i.e. almost free electrons, then the equal energy surfaces are approximately spherical and that with the largest energy has a radius equal to the semi-diagonal of the first zone which corresponds to the wave number k_{max} where $k_{\text{max}} = \sqrt{3}\,\pi/a$. Interpreting this in terms of energy, we have for an electron, the kinetic energy

$$E = \tfrac{1}{2}\,mv^2 = \frac{p^2}{2m}$$

where p is its momentum, and putting p in terms of the de Broglie wavelength λ we have

$$E = \frac{1}{2m}\,\frac{\mathbf{h}^2}{\lambda^2} = \frac{\mathbf{h}^2 k^2}{8\pi^2 m}$$

since $k = 2\pi/\lambda$. Substituting for k_{max} we have

$$E_{\text{max}} = \frac{\mathbf{h}^2}{8\pi^2 m}\cdot\frac{3\pi^2}{a^2} = \frac{3\mathbf{h}^2}{8a^2 m}$$

$$= \frac{3 \times (6{\cdot}62 \times 10^{-27})^2}{8 \times (3 \times 10^{-8})^2 \times 9{\cdot}11 \times 10^{-28} \times 1{\cdot}6 \times 10^{-12}}\ \text{eV}$$

$$= \underline{12{\cdot}5\ \text{eV}}$$

Problem 4.13

State the assumptions on which the Fermi-Dirac distribution law is based and quote the distribution function. Sketch graphs to display the function at absolute zero and at another temperature, and briefly explain their forms.

Derive an expression for the Fermi level at absolute zero, using the free electron model. Evaluate the value for sodium using the following data—

Density of sodium	$= 0{\cdot}971$ g cm^{-3}
Atomic wt. of sodium	$= 22{\cdot}99$
Avogadro's number	$= 6{\cdot}025 \times 10^{23}$ mole^{-1}
Electronic mass	$= 9{\cdot}107 \times 10^{-28}$ g
Electronic charge	$= 1{\cdot}601 \times 10^{-19}$ C
Planck's constant	$= 6{\cdot}624 \times 10^{-27}$ erg sec

Outline the explanation of the insignificant contribution of the conduction electrons to the specific heat of a metallic element at room temperature.

[B.Sc. (Special), Pt. II, London, 1962]

Ans. 3·16 eV

FREE CHARGE CARRIERS IN METALS AND SEMICONDUCTORS

Problem 4.14

Show that the electrical conductivity σ of a metal may be expressed as $\dfrac{n\mathbf{e}^2\tau}{m}$, where **e** is the electronic charge, n the number of electrons per unit volume and τ the relaxation time of the electrons. How is this expression for electrical conductivity modified for an intrinsic semiconductor?

The intrinsic carrier density of germanium at 27°C is $2{\cdot}40 \times 10^{19}$ m^{-3}. Calculate the intrinsic resistivity if the electron and hole mobilities are respectively 0·38 m^2 volt^{-1} sec^{-1}, and 0·18 m^2 volt^{-1} sec^{-1}.

[Dip. Tech. Elec. Eng., Salford, 1963]

SOLUTION

The conductivity σ of a substance containing n charge carriers per unit volume, each of mobility μ and charge **e** may be expressed as

$$\sigma = n\mathbf{e}\mu$$

When both holes and electrons take part in the conduction process the resultant conductivity is the sum of both contributions, thus

$$\sigma = \sigma_h + \sigma_e = n_h\mathbf{e}\mu_h + n_e\mathbf{e}\mu_e$$

where the subscripts h and e refer to holes and electrons respectively. In an intrinsic semiconductor, $n_h = n_e = n_i$, where n_i is the intrinsic charge carrier density. Thus for the intrinsic semiconductor

$$\begin{aligned}\sigma &= n_i \mathbf{e} (\mu_h + \mu_e) \\ &= 2{\cdot}40 \times 10^{19} \times 1{\cdot}60 \times 10^{-19}(0{\cdot}38 + 0{\cdot}18) \\ &= 2{\cdot}15 \text{ ohm}^{-1} m^{-1}\end{aligned}$$

or the resistivity $\rho = \dfrac{1}{2{\cdot}15} = \underline{0{\cdot}465 \text{ ohm.m}}$

Problem 4.15

Compare and contrast the energy band structures of metals and semiconductors. Indicate a method by which the energy gap in an intrinsic semiconductor may be evaluated.

Calculate the charge carrier relaxation time, at room temperature, in an n-type silicon specimen for which the electrical conductivity is 500 mho m^{-1} and the concentration of donor impurities is 10^{22} m^{-3}. You may assume an electron effective mass of $0{\cdot}5\ \mathbf{m}_e$.

[B.Sc. (General Hons.), Durham, 1962]

SOLUTION

It may be assumed that all the donors are ionised at room temperature and that intrinsic conduction is negligible, thus the charge carrier density $n = 10^{22}$ m^{-3}.

The conductivity σ of the material is related to the relaxation time τ of the carrier electrons by the expression

$$\sigma = \frac{n\mathbf{e}^2\tau}{\mathbf{m}_e}$$

where $\mathbf{m}_e$ is the effective mass of the carriers. Thus

$$\begin{aligned}\tau &= \frac{\mathbf{m}_e\sigma}{n\mathbf{e}^2} \\ &= \frac{0{\cdot}5 \times 9{\cdot}11 \times 10^{-31} \times 500}{10^{22} \times (1{\cdot}6 \times 10^{-19})^2} = \underline{8{\cdot}9 \times 10^{-13} \text{ sec}}\end{aligned}$$

Problem 4.16

Describe an experiment to determine the intrinsic energy gap of germanium.

Explain how the measurement of the Hall coefficient of a semiconductor containing only one type of carrier gives information about the number of carriers present. The dimensions of a crystal of such a material are 1 mm × 1 cm × 2 cm, and a current of 10 mA is passed between the two smallest faces. A magnetic flux density of 0·5 Wb m^{-2} is applied parallel to the 1 mm edges, and this produces a Hall voltage of 1·86 millivolts. Determine the carrier density.

[Dip. Tech. Elec. Eng., Salford, 1962]

SOLUTION

The Hall coefficient R_H of a material is defined as $R_H = E_H/BJ$, where the Hall electric field E_H is produced perpendicular to both the flux density B and the current density J, (i.e. across the 1 cm dimension). Also, if the velocity distribution of the electrons is neglected, $R_H = 1/n\mathbf{e}$ where n is the charge density and $\mathbf{e}$ is the electronic charge. Thus

$$\frac{1}{n\mathbf{e}} = \frac{E_H}{BJ}$$

or

$$n = \frac{BJ}{E_H\mathbf{e}}$$

Substitution gives

$$n = \frac{0{\cdot}5 \times \dfrac{10 \times 10^{-3}}{10^{-2} \times 10^{-3}}}{\dfrac{1{\cdot}86 \times 10^{-3}}{10^{-2}} \times 1{\cdot}60 \times 10^{-19}} = \underline{1{\cdot}68 \times 10^{22}\ \text{m}^{-3}}$$

Problem 4.17

What is the Hall effect and how is it explained?

Compare the experimental values of the Hall coefficient, R, for copper and zinc, viz. $-5{\cdot}5 \times 10^{-4}$ e.m.u. and $+3{\cdot}3 \times 10^{-4}$ e.m.u. respectively, with the values predicted on the free electron theory of metals, if all the valence electrons are assumed free.

(The charge carried by a gram equivalent of ions is 9,650 e.m.u.

At. wts. Cu, 63·6; Zn, 65·4
Densities Cu, 8·89; Zn, 7·1)

[B.Sc. (Honours), Liverpool, 1962]

SOLUTION

The theoretical values of the Hall coefficients may be determined as follows: 63·6 g of copper contain N_0 (Avogadro's number) atoms. Therefore 1 cc contains $N_0/63{\cdot}6 \times \rho$ atoms where ρ is the density of copper. Since copper is monovalent each atom provides one free electron, thus the free electron density $n = N_0 \times 8{\cdot}89/63{\cdot}6$ cm^{-3}.

Neglecting the velocity distribution of the free electrons the Hall coefficient

$$R = -\frac{1}{n\mathbf{e}}$$

where **e** is the electronic charge, i.e.

$$R = -\frac{63{\cdot}6}{8{\cdot}89 \times N_0\mathbf{e}}\ \text{cm}^3\ \text{e.m.u.}^{-1}$$

But $N_0\mathbf{e} = 9{,}650$ e.m.u., therefore

$$R = -\frac{63{\cdot}6}{8{\cdot}89 \times 9{,}650} = \underline{-\ 7{\cdot}4 \times 10^{-4}\ \text{cm}^3\ \text{e.m.u.}^{-1}}$$

Similarly, for zinc which provides 2 free electrons per atom we have

$$R = -\frac{65{\cdot}4}{2 \times 7{\cdot}1 \times 9{,}650} = \underline{-\ 4{\cdot}77 \times 10^{-4}\ \text{cm}^3\ \text{e.m.u.}^{-1}}$$

Problem 4.18

Describe the different types of chemical binding which take place in semiconductors, taking as examples, Ge, In, Sb, and PbS. How does the type of binding affect the crystal structure? Discuss in terms of chemical bonds the concepts: conduction electron, positive hole, donor impurity and acceptor impurity.

A crystal of pure Ge has sufficient Sb added so as to produce $1{\cdot}5 \times 10^{16}$ Sb atoms per cm^3. Assuming that the Sb atoms enter the crystal substitutionally and are all ionised, calculate approximately the number of conduction electrons and positive holes per cm^3 of the crystal. Show that the positive holes make a negligible contribution to the conductivity and calculate its value at room temperature.

Sufficient In is now added to produce $2{\cdot}5 \times 10^{16}$ In atoms per cm^3. Assuming that the In atoms also enter the crystal substitutionally, that all accept an electron and that no Sb is lost in the process, calculate the number of conduction electrons and holes per cm^3 of the crystal. Show that now the conduction electrons make a negligible contribution to the conductivity and calculate its value at room temperature. (The concentration of conduction electrons and holes in pure Ge at room temperature may be taken as $2{\cdot}5 \times 10^{13}$ cm^{-3}. The mobility of electrons in Ge at room temperature may be taken as 3,800 cm^2 $volt^{-1}$ sec^{-1} and the mobility of holes as 1,800 cm^2 $volt^{-1}$ sec^{-1}.)

[B.Sc. (Special), Sheffield, 1962]

SOLUTION

Initially the conduction electron density, n_e, equals the density of the donor (Sb) atoms, i.e.

$$n_e = 1{\cdot}5 \times 10^{16}\ cm^{-3}$$

The density of the minority carriers (holes), n_h, can be found from the relation

$$n_e n_h = n_i^2$$

where n_i is the density of the intrinsic carriers at the temperature considered. Thus

$$n_h = \frac{n_i^2}{n_e} = \frac{(2{\cdot}5 \times 10^{13})^2}{1{\cdot}5 \times 10^{16}} = 4{\cdot}16 \times 10^{10}\ cm^{-3}$$

As explained in Problem 4.14 the conductivity σ of the material is given by

$$\begin{aligned}\sigma &= n_e\mu_e\,\mathbf{e} + n_h\mu_h\,\mathbf{e}\\ &= (1{\cdot}5 \times 10^{16} \times 3{,}800 + 4{\cdot}16 \times 10^{10} \times 1{,}800)1{\cdot}6 \times 10^{-19}\\ &= \underline{9{\cdot}12\ ohm^{-1}\ cm^{-1}}\end{aligned}$$

and the contribution due to the holes is clearly negligible.

The first $1{\cdot}5 \times 10^{16}\ cm^{-3}$ In atoms added to the germanium accept the $1{\cdot}5 \times 10^{16}$ electrons donated by the Sb. The remaining

$1{\cdot}0 \times 10^{16}\ \text{cm}^{-3}$ In atoms accept electrons from the valence band so producing $1{\cdot}0 \times 10^{16}$ holes per cm^3, i.e. $n_h = 1{\cdot}0 \times 10^{16}\ \text{cm}^{-3}$.

Proceeding as before to find the concentration of the minority carriers (now electrons), we have

$$n_e = \frac{n_i^2}{n_h} = \frac{(2{\cdot}5 \times 10^{13})^2}{1{\cdot}0 \times 10^{16}} = 6{\cdot}25 \times 10^{10}\ \text{cm}^{-3}$$

The conductivity σ is given by

$$\begin{aligned}\sigma &= n_e\mu_e\mathbf{e} + n_h\mu_h\mathbf{e} \\ &= (6{\cdot}25 \times 10^{10} \times 3{,}800 + 1{\cdot}0 \times 10^{16} \times 1{,}800)1{\cdot}6 \times 10^{-19} \\ &= \underline{2{\cdot}88\ \text{ohm}^{-1}\ \text{cm}^{-1}}\end{aligned}$$

and again the minority carriers contribute negligibly to the conductivity.

Problem 4.19

Explain the difference between n and p type germanium and explain how each type of material may be made from impure (~99%) germanium.

If there are 5×10^{22} donors per m^3, each containing one electron, in a piece of semiconductor and the mobility of electrons in the conduction band is $0{\cdot}40\ \text{m}^2\ \text{volt}^{-1}\ \text{sec}^{-1}$, what is the conductivity of the material in the condition of carrier exhaustion?

Intrinsic conduction may be neglected.

[B.Sc. (General Hons.), Durham, 1961]

Ans. $3{\cdot}2 \times 10\ \text{ohm}^{-1}\ \text{m}^{-1}$

Problem 4.20

Discuss the electrical conductivity and its temperature dependence in intrinsic and extrinsic semi-conductors.

A specimen of n-type germanium has a Hall coefficient of $-62{\cdot}5\ \text{coulomb}^{-1}\ \text{cm}^3$ at 290°K. Indium is now diffused into the specimen and the Hall coefficient is found to have changed to $-68{\cdot}0\ \text{coulomb}^{-1}\ \text{cm}^3$. Find the density of the impurity centres if it can be assumed that the indium is distributed uniformly and that it has entered the lattice substitutionally. Calculate also the changes in positions of the Fermi level and illustrate your answer by band diagrams.

Effective density of states near edges at 290°K = 2×10^{19} cm^{-3}.

[B.Sc. (Special), Pt. II, Leicester, 1963]

Ans. 10^{16}/cm^3; Fermi level is lowered from 0·127 eV to 0·130 eV below the conduction band

Problem 4.21

Explain carefully the concept of a hole and show with the aid of some diagrams that the Hall effect is able to give the sign of the charge carriers in a conductor.

In n-type germanium the donor ionisation energy is approximately 0·01 eV and the energy gap, E, in germanium is approximately 0·7 eV. Indicate how the number of charge carriers will vary with temperature. Assume that the density of impurities is 10^{13} cm^{-3} and that the density of electrons in the conduction band extrapolated to high temperature is 10^{19} cm^{-3}. How will the inverse Hall coefficient vary with temperature?

[B.Sc. (Honours), Manchester, 1963]

Ans. The inverse Hall coefficient is proportional to the number of charge carriers which at selected temperatures 5·8°K, 405°K and 810°K are $4{\cdot}54 \times 10^{11}$ cm^{-3}, $4{\cdot}54 \times 10^{14}$ cm^{-3} and $6{\cdot}74 \times 10^{16}$ cm^{-3} respectively. Also, in the temperature range 50°K to the temperature at which intrinsic conduction becomes significant the charge carrier concentration is constant at 10^{13} cm^{-3}.

MISCELLANEOUS

Problem 4.22

State Richardson's equation and define each symbol appearing in it.

Discuss the phenomenon of thermionic emission.

In a power diode the cathode consists of a tungsten strip filament of work function 4·50 V. If the maximum temperature-limited current is 10 mA at a filament temperature of 2,000°K, find the maximum current at a filament temperature of 2,600°K.

[Dip. Tech. Elec. Eng., Salford, 1962]

SOLUTION

From Richardson's equation we can write that

$$I_{\text{sat}} \propto T^2 \exp\left(-\frac{\mathbf{e}\phi}{\mathbf{k}T}\right)$$

where I_{sat} is the maximum thermionic current at temperature T, ϕ is the work function of the surface, **e** is the electronic charge and **k** is Boltzmann's constant. Thus

$$I_{\text{sat}}^{2{,}600} = I_{\text{sat}}^{2{,}000}\left(\frac{2{,}600}{2{,}000}\right)^2 \frac{\exp\left(\dfrac{-4\cdot 5\ \mathbf{e}}{2{,}600\ \mathbf{k}}\right)}{\exp\left(\dfrac{-4\cdot 5\ \mathbf{e}}{2{,}000\ \mathbf{k}}\right)}$$

where the superscripts refer to the temperatures for the respective saturation currents. Therefore

$$I_{\text{sat}}^{2{,}600} = 10 \times \left(\frac{2{,}600}{2{,}000}\right)^2 \times \exp\left(\frac{4\cdot 5 \times 1\cdot 6 \times 10^{-19} \times 600}{1\cdot 38 \times 10^{-23} \times 2{,}000 \times 2{,}600}\right) \text{mA}$$

$$= \underline{7\cdot 03 \text{ mA}}$$

Problem 4.23

Describe (*a*) how the concept of dislocations in crystals may account for a large amount of slip in a crystal plane, (*b*) one way in which dislocations give an explanation of work hardening.

Etch pits at a low angle grain boundary are found to be 10^{-4} cm apart. If the effective atomic spacing is 3×10^{-8} cm determine the angle between the grains.

[H.N.C. Applied Physics, Salford, 1963]

SOLUTION

The angle between the grains $\alpha = b/d$ where b is the Burger's vector of the dislocation and d is the distance between the dislocations. b is identified with the atomic spacing and d with the separation of the etch pits. Thus

$$\alpha = \frac{3 \times 10^{-8}}{10^{-4}} = \underline{3 \times 10^{-4}\ \text{rad}}$$

5
Wave Mechanics

DUALISM

Problem 5.1

Describe how the Compton scattering of X-rays has been investigated experimentally.

Give an explanation of the 'modified' and 'unmodified' wavelengths found in the scattered radiation, giving the necessary theory.

Monochromatic X-rays of wavelength $\lambda = 0{\cdot}124$ Å are scattered from a carbon block. Determine the wavelength of the X-rays scattered (*a*) through 45°, and (*b*) through an angle such that the recoil electron has the maximum possible kinetic energy.

[Dip. Tech. Physics, Pt. I, Salford, 1961]

SOLUTION

(*a*) The change in wavelength of the X-rays, $\lambda' - \lambda$, due to the scattering by a free electron is

$$\lambda' - \lambda = \frac{\mathbf{h}}{mc}(1 - \cos\phi)$$

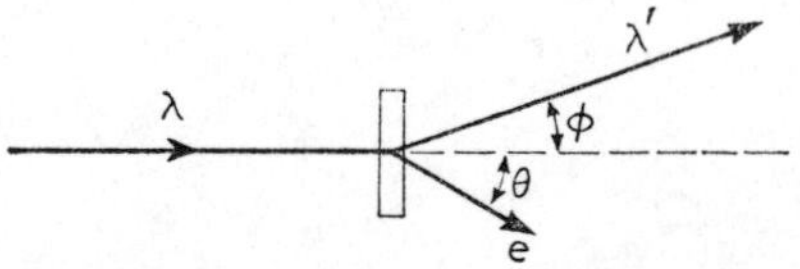

Fig. 5.1

where ϕ is the angle through which the radiation is scattered (Fig. 5.1), and the other symbols have their usual meaning.

The Compton wavelength

$$\frac{\mathbf{h}}{m\mathbf{c}} = \frac{6{\cdot}625 \times 10^{-27}}{9{\cdot}11 \times 10^{-28} \times 3 \times 10^{10}}$$

$$\therefore \qquad \lambda' - \lambda = \frac{6{\cdot}625 \times 10^{-9}}{9{\cdot}11 \times 3}[1 - \cos 45°] = 0{\cdot}007\ \text{Å}$$

$$\therefore \qquad \lambda' = \lambda + 0{\cdot}007 = 0{\cdot}124 + 0{\cdot}007 = \underline{0{\cdot}131\ \text{Å}}$$

(*b*) The electron has maximum recoil energy when the radiation is scattered through 180° (i.e. $\cos \phi = -1$).

$$\lambda' - \lambda = \frac{6{\cdot}625}{9{\cdot}11 \times 3} \times 10^{-9} \times 2 = 0.0485$$

$$\lambda' = 0{\cdot}124 + 0{\cdot}0485 = \underline{0{\cdot}173\ \text{Å}}$$

Problem 5.2

Outline the theory of the Compton effect. How well is this theory substantiated by experiments?

The K_α radiation from a molybdenum target, $\lambda = 0{\cdot}708$ Å is scattered from a carbon block. Calculate the change in wavelength ($\Delta\lambda = \mathbf{h}(1 - \cos \phi)/m\mathbf{c}$) of the X-rays scattered through 180° and determine the maximum kinetic energy in eV of the recoil electrons produced by this scattering process.

[B.Sc. (Special), Sheffield, 1962]

SOLUTION

The formula for the change in wavelength is given in the question

$$\Delta\lambda = \frac{\mathbf{h}}{m\mathbf{c}}(1 - \cos \phi)$$

where **h** is Planck's constant, m is the electron mass and **c** the velocity of light. The angle through which the X-rays are scattered is 180° in this case, so that

$$\Delta\lambda = \frac{\mathbf{h}}{m\mathbf{c}}(1 - \cos \pi) = \frac{2\mathbf{h}}{m\mathbf{c}}$$

Therefore

$$\Delta\lambda = \frac{2 \times 6{\cdot}62 \times 10^{-27}}{9{\cdot}11 \times 10^{-28} \times 3 \times 10^{10}} = \underline{0{\cdot}0484\ \text{Å}}$$

The change in wavelength is such that the wavelength of the scattered radiation is increased, thus the wavelength of the scattered radiation = 0·756 Å.

The k.e. of the electron is $\mathbf{h}(\nu - \nu')$ where ν and ν' are the frequencies before and after collision respectively. Therefore

$$\text{k.e. of electron} = \mathbf{hc}\left(\frac{1}{\lambda} - \frac{1}{\lambda'}\right) = \frac{\mathbf{hc}}{\lambda\lambda'}(\lambda' - \lambda)$$

$$= \frac{6{\cdot}62 \times 10^{-27} \times 3 \times 10^{10}}{10^{-8}} \times \left(\frac{0{\cdot}0484}{0{\cdot}708 \times 0{\cdot}756}\right)$$

$$= 1{\cdot}796 \times 10^{-9}\ \text{ergs}$$

or since 1 keV = $1{\cdot}6 \times 10^{-9}$ ergs

$$\text{k.e. of electron} = \underline{1{\cdot}122\ \text{keV}}$$

Problem 5.3

Give an account of the experimental evidence which demonstrates the wave-like properties of moving electrons.

Discuss the interpretation of these 'electron waves.'

Calculate the de Broglie wavelength of waves associated with a beam of neutrons of energy 0·025 eV.

[Dip. Tech. Physics, Pt. I, Salford, 1961]

SOLUTION

The de Broglie wavelength associated with a particle of mass m moving with a velocity v is

$$\lambda = \frac{\mathbf{h}}{mv}$$

If the energy associated with such a particle is E, then

$$E = \tfrac{1}{2}mv^2$$

and $$mv = \sqrt{(2mE)}$$

$$\therefore \qquad \lambda = \frac{\mathbf{h}}{\sqrt{(2mE)}}$$

In this case

$$E = 0{\cdot}025 \text{ eV} = 0{\cdot}025 \times 1{\cdot}6 \times 10^{-12}$$
$$= 0{\cdot}4 \times 10^{-13} \text{ ergs}$$

Also the mass of the neutron $m = 1{\cdot}67 \times 10^{-24}$ g. Therefore

$$\lambda = \frac{6{\cdot}62 \times 10^{-27}}{2 \times 1{\cdot}67 \times 10^{-24} \times 0{\cdot}4 \times 10^{-13}} = 1{\cdot}81 \times 10^{-8} \text{ cm}$$
$$= \underline{1{\cdot}81 \text{ Å}}$$

Problem 5.4

Describe the experiments of Davisson and Germer which demonstrated the wave-like properties of beams of electrons.

A stream of electrons of energy 250 eV is incident on the surface of a platinum foil at an angle of 30° to the normal. The change in potential of an electron entering the metal is 12 eV. Calculate:

(*a*) The refractive index of the platinum for the electrons of this energy.
(*b*) The wave velocity of the electron waves in the metal, and the actual velocity of the electrons.

[B.Sc. (Special), Pt. I, Bristol, 1961]

SOLUTION

(*a*) Let the velocity of the electrons outside the platinum be v_1 and inside, v_2, and let the potential barrier encountered by the electron be V (Fig. 5.2). By the law of conservation of energy

$$\tfrac{1}{2}mv_1^2 = \tfrac{1}{2}mv_2^2 - V$$

$$\therefore \qquad v_1^2 = v_2^2 - \frac{2V}{m}$$

$$v_2 = v_1\sqrt{\left(1 + \frac{V}{\tfrac{1}{2}mv_1^2}\right)}$$

The refractive index

$$\mu = \frac{v_2}{v_1} = \sqrt{\left(1 + \frac{V}{\frac{1}{2}mv_1^2}\right)}$$

Both the potential and k.e. of the electron are given in electron volts, i.e. $V = 12$ eV and $\frac{1}{2}mv_1^2 = 250$ eV. Therefore

$$\mu = \sqrt{1 + \frac{12}{250}} = \underline{1{\cdot}023}$$

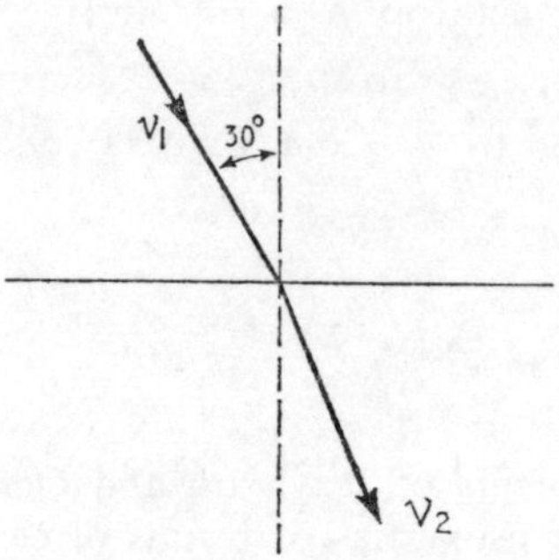

Fig. 5.2

(*b*) The velocity of the electrons in the metal $v_2 = 1{\cdot}023\ v_1$. Therefore

$$v_2 = 1{\cdot}023 \sqrt{\left(\frac{2\mathbf{e}V_1}{m}\right)}$$

where $\mathbf{e}V_1$ is the energy of the electron beam in free space. Now $V_1 = 250$ volts $= \frac{5}{6}$ e.s. volts. Therefore

$$v_2 = 1{\cdot}023 \left(\frac{2 \times 4{\cdot}8 \times 10^{-10} \times 5}{9{\cdot}11 \times 10^{-28} \times 6}\right)^{\frac{1}{2}} = \underline{0{\cdot}959 \times 10^9 \text{ cm sec}^{-1}}$$

The wave velocity of the electron waves ω is given by

$$\omega = \frac{\mathbf{c}^2}{v^2} = \frac{9 \times 10^{20}}{0{\cdot}959 \times 10^9} = \underline{9{\cdot}38 \times 10^{11} \text{ cm sec}^{-1}}$$

Problem 5.5

Give a concise account of the experimental evidence for the wave-particle dualism of electrons and electromagnetic radiation and discuss the reconciliation of the wave and particle treatments of these concepts.

Electrons of 54 eV energy are incident, at an angle of 25° with the normal, on a set of lattice planes whose distance apart is 0·91 Å, and experience a Bragg reflection of the first order.

Calculate the wavelength of the associated electron waves and compare it with the de Broglie value.

[Dip. Tech. Physics, Pt. I, Salford, 1963]

SOLUTION

First order Bragg reflection occurs, when the following condition applies

$$\lambda = 2d \sin \theta$$

where λ is the wavelength associated with the electron, d is the

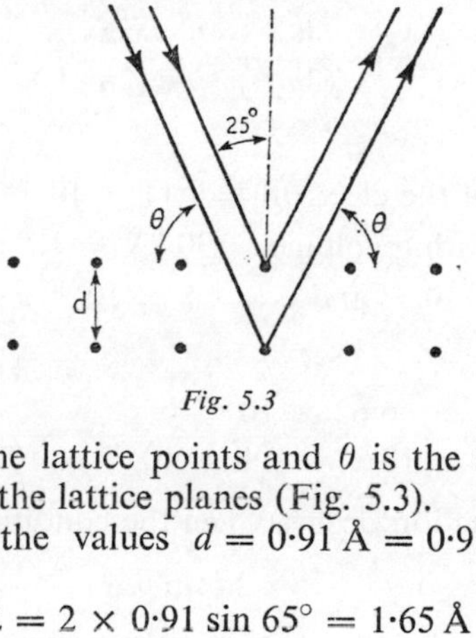

Fig. 5.3

distance between the lattice points and θ is the angle between the incident beam and the lattice planes (Fig. 5.3).

Substitution of the values $d = 0{\cdot}91$ Å $= 0{\cdot}91 \times 10^{-8}$ cm, and $\theta = 65°$ gives

$$\lambda = 2 \times 0{\cdot}91 \sin 65° = \underline{1{\cdot}65 \text{ Å}}$$

Consider the de Broglie wavelength associated with these electrons

$$\lambda = \frac{\mathbf{h}}{mv} = \frac{\mathbf{h}}{\sqrt{(2mE)}}$$

where E is the energy of the electron beam. Now $E = 54$ eV $= 54 \times 1{\cdot}6 \times 10^{-12}$ ergs and $m = 9{\cdot}11 \times 10^{-28}$ g. Therefore

$$\lambda = \frac{6{\cdot}62 \times 10^{-27}}{\sqrt{(2 \times 9{\cdot}11 \times 10^{-28} \times 54 \times 1{\cdot}6 \times 10^{-12})}}$$
$$= 1{\cdot}66 \times 10^{-8} \text{ cm} = \underline{1{\cdot}66 \text{ Å}}$$

Problem 5.6

By using the relation $E = \mathbf{h}\omega$ for the frequency of an electron wave, derive de Broglie's relation for the wavelength λ expressed in terms of the momentum p. Derive, and comment upon, the formulae for the phase and group velocities of the electron waves.

Electrons from a heated filament are accelerated by a potential of 30 kV, collimated and transmitted through a thin sheet of aluminium. Calculate the angle of the first-order diffraction pattern.

(Lattice constant of aluminium may be taken as 4·05 Å.)

[B.Sc. (Ordinary), Bristol, 1963]

SOLUTION

The wavelength associated with an electron beam of energy E is given by the formula

$$\lambda = \frac{\mathbf{h}}{\sqrt{(2mE)}} = \frac{\mathbf{h}}{\sqrt{(2m\mathbf{e}V)}}$$

where

m = mass of the electron = $9{\cdot}11 \times 10^{-28}$ g
V = accelerating voltage = 30 kV = 10^2 e.s.u. volts
$\mathbf{h} = 6{\cdot}62 \times 10^{-27}$ and $\mathbf{e} = 4{\cdot}8 \times 10^{-10}$ e.s.u.

Therefore

$$\lambda = \frac{6{\cdot}62 \times 10^{-27}}{(2 \times 9{\cdot}11 \times 10^{-28} \times 4{\cdot}8 \times 10^{-10} \times 10^2)^{\frac{1}{2}}} = 0{\cdot}0708 \text{ Å}$$

First order diffraction occurs when the condition

$$\lambda = 2d \sin\theta \qquad (5.6.1)$$

is satisfied. d is the distance between the lattice planes and θ is the angle the incident beam makes with these planes (Fig. 5.4.), i.e. the angle between incident and emergent beams is 2θ. Rewriting Equation 5.6.1,

$$\sin\theta = \frac{\lambda}{2d}$$

If θ is to be a minimum then d must be a maximum. Therefore

$$\sin\theta_{\text{min}} = \frac{\lambda}{2d_{\text{max}}}$$

Now $$d = \frac{a_0}{(h^2 + k^2 + l^2)^{\frac{1}{2}}}$$

where a_0 is the lattice constant and h, k, l are the Miller indices. Aluminium is face centred cubic and therefore

$$d_{max} = \frac{a_0}{\sqrt{3}} = \frac{4{\cdot}05}{\sqrt{3}}\,\text{Å}$$

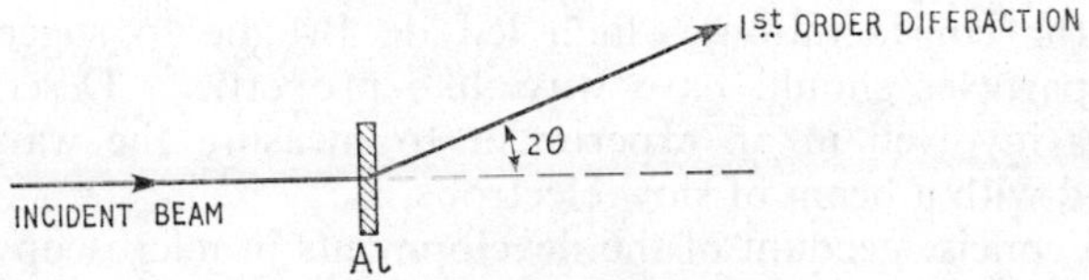

Fig. 5.4

$$\therefore \qquad \sin\theta_{min} = \frac{0{\cdot}0708 \times 53}{2 \times 4{\cdot}05}$$

$$\theta_{min} = 46'$$

Therefore, the angle of the 1st order diffraction pattern $2\theta = \underline{1° \; 32'}$.

Problem 5.7

Describe an experiment which measures the changes in wavelength which occur when X-rays are scattered by matter and give a theory which accounts for this phenomenon.

X-rays of wavelength 0·708 Å are scattered by a block of carbon. Calculate the change in wavelength in the radiation scattered through 60°.

[B.Sc. (Special), Pt. I, London, 1959]

Ans. 0·0122 Å

Problem 5.8

Calculate the r.m.s. velocity of thermal neutrons at 300°K.

What is the wavelength of the de Broglie waves associated with a homogeneous beam of neutrons travelling with this velocity?

Mention two consequences of the order of magnitude of this wavelength.

[B.Sc. Normal, Pt. II, III, Liverpool, 1962]

Ans. velocity = $2{\cdot}73 \times 10^5$ cm/sec; 1·452 Å

Problem 5.9

Outline the considerations which led de Broglie to suggest that moving particles should have wave-like properties. Describe the principles involved in an experiment to measure the wavelength associated with a beam of slow electrons.

Give a concise account of the developments in microscopy which have resulted from the wave-like properties of the electron and discuss the limitations at present existing.

Suppose a microscope using X-rays could be constructed, what would be the photon energy required to give a resolution equal to the theoretical limit of an electron microscope using 50 keV electrons? (Neglect relativistic effects.)

[Dip. Tech. Physics, Pt. I, Salford, 1962]

Ans. 0·226 MeV

Problem 5.10

Discuss briefly evidence for the wave nature and corpuscular nature of matter and light.

Derive an expression for the de Broglie wavelength of electrons which have been accelerated from rest through a potential difference V, and hence show that a region of space of positive potential V_0 has a refractive index

$$\mu = \sqrt{\left(1 + \frac{V_0}{V}\right)}$$

for such a beam.

A parallel beam of 40 kV electrons is incident normally on a thin parallel-sided film of material inside which the average potential is 10 volts positive with respect to the outside. Calculate the thickness of the film if it introduces a phase change of $\pi/2$ in the electron-wave front.

[Nat. Science Tripos, Pt. I, Cambridge, 1962]

Ans. 122·6 Å

HEISENBERG UNCERTAINTY PRINCIPLE

Problem 5.11

Explain the significance of Heisenberg's uncertainty principle. Illustrate your explanation with an example showing how the attainable precision of physical measurement is controlled by the principle.

A particle is placed in a potential well of width 'a' having infinitely high walls. Estimate the energy of the lowest level in the well.

[B.Sc. Higher, Pt. II, III, Liverpool, 1963]

SOLUTION

The uncertainty principle states that

$$\Delta x \,.\, \Delta p \sim \mathbf{h}$$

where Δx is the uncertainty in the position of the particle and Δp is the uncertainty in the particle's momentum. This is true only if the measurement of x and p are carried out simultaneously. ($\mathbf{h}$ is Planck's constant.)

In this particular case the particle has no chance of escaping the infinitely high potential well and so the particle is localised to within $\Delta x = a$. Therefore

$$\Delta p \sim \frac{\mathbf{h}}{a}$$

The kinetic energy of the lowest level must at least be $\sim (\Delta p)^2/2m$. Therefore the energy of the particle in the lowest level $\sim \underline{\mathbf{h}^2/2ma^2}$.

Problem 5.12

State Heisenberg's uncertainty principle and illustrate it by reference to a few idealised experiments.

Classical physics requires a one-dimensional oscillator, in its ground state, to be at rest, i.e $\delta x = 0$ and $\delta p = 0$

If in the quantum mechanical treatment we must have $\delta x \neq 0$, $\delta p \neq 0$ and the total energy of a harmonic oscillator is

$$E = \frac{p^2}{2m} + \frac{kx^2}{2}$$

where k is a force constant, then, by assuming that $p \sim \delta p$ and $x \sim \delta x$, show that the minimum value of E (by differentiation or otherwise) is

$$E \sim \mathbf{h}\sqrt{\left(\frac{k}{m}\right)}$$

Comment on this result.

[Dip. Tech. Physics, Pt. I, Salford, 1964]

SOLUTION

The total energy of the oscillator is

$$E = \frac{p^2}{2m} + \frac{1}{2}kx^2$$

$$\therefore \qquad E = \frac{(\delta p)^2}{2m} + \frac{k(\delta x)^2}{2} \qquad (5.12.1)$$

if it is assumed that $p \sim \delta p$ and $x \sim \delta x$.

According to the uncertainty principle $\delta p \,.\, \delta x = h$ and substituting for δp in terms of δx in Equation 5.12.1

$$E = \frac{h^2}{2m(\delta x)^2} + \frac{k(\delta x)^2}{2} \qquad (5.12.2)$$

The minimum value of E occurs for $\mathrm{d}E/\mathrm{d}(\delta x) = 0$, i.e. when

$$-\frac{\mathbf{h}^2}{m(\delta x)^3} + k(\delta x) = 0$$

$$\therefore \qquad (\delta x)^2 = \frac{\mathbf{h}}{\sqrt{(km)}}$$

Substitution of this value of $(\delta x)^2$ into Equation 5.12.2 gives,

$$E = \frac{\mathbf{h}^2}{2m}\frac{\sqrt{(km)}}{\mathbf{h}} + \frac{k}{2}\frac{\mathbf{h}}{\sqrt{(km)}}$$

$$\therefore \qquad \underline{E \sim \mathbf{h}\sqrt{\left(\frac{k}{m}\right)}}$$

Problem 5.13

State Heisenberg's uncertainty principle and illustrate it by reference to an idealised experiment.

A 1-eV electron beam falls normally upon a narrow slit of width 10^{-3} cm. Calculate the width of the diffraction maximum at a distance of 5 m.

[B.Sc. Applied Science, Pt. I, Durham, 1962]

Ans. 1·226 mm

THE SCHRÖDINGER WAVE EQUATION

Problem 5.14

A beam of electrons, each of velocity v, constitutes a current of 1 μA flowing parallel to the x-axis. An abrupt potential step at $x = 0$ reduces the velocity of the electrons proceeding beyond the step to 0·25 v. Deduce, using Schrödinger's equation, the current beyond the potential barrier.

Discuss briefly the non-Newtonian behaviour of electrons in solids.

[B.Sc. Tech. (Honours), Manchester, 1963]

SOLUTION

Consider the electron beam to have an energy E, and the potential step it encounters at $x = 0$ to be of height V. E must be greater than

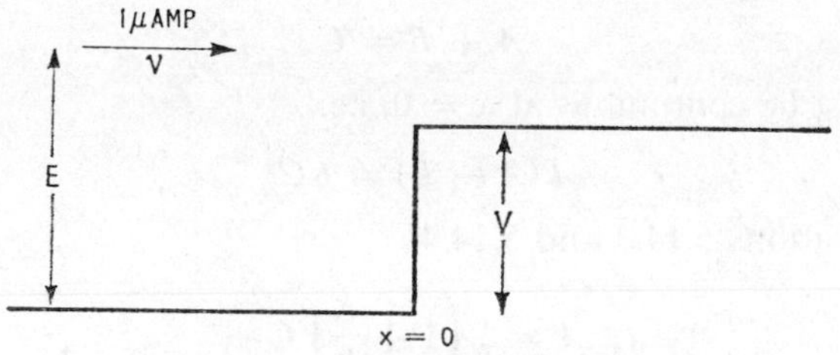

Fig. 5.5

V in order for transmission to occur (Fig. 5.5). The Schrödinger equation is

$$\frac{d^2\psi}{dx^2} + \frac{2m}{\hbar^2}(E - V)\psi = 0$$

For the region $x < 0$, V is zero. Therefore

$$\frac{d^2\psi}{dx^2} + \frac{2mE}{\hbar^2}\psi = 0$$

$$\frac{d^2\psi}{dx^2} + k^2\psi = 0 \qquad (5.14.1)$$

where $k^2 = 2mE/\hbar^2$. The solution of Equation 5.14.1 gives

$$\psi = A \exp(ikx) + B \exp(-ikx)$$

where $A \exp(ikx)$ represents a beam travelling in the $+ve$ x direction and $B \exp(-ikx)$ represents a reflected beam travelling in the $-ve$ x direction.

For the region $x > 0$, the Schrodinger equation is

$$\frac{d^2\psi}{dx^2} + \frac{2m(E - V)}{\hbar^2}\psi = 0$$

$$\therefore \qquad \frac{d^2\psi}{dx^2} + K^2\psi = 0 \qquad (5.14.2)$$

where $K^2 = 2m(E - V)/\hbar^2$. The solution of Equation 5.14.2 gives

$$\psi = C \exp(+iKx)$$

The solution $\exp(-iKx)$ is rejected as having no physical meaning.

Applying boundary conditions, ψ must be continuous at $x = 0$, i.e.

$$A + B = C \qquad (5.14.3)$$

$d\psi/dx$ must be continuous at $x = 0$, i.e.

$$k(A - B) = KC \qquad (5.14.4)$$

From Equations 5.14.3 and 5.14.4

$$A = \tfrac{1}{2}\left(1 + \frac{K}{k}\right) C$$

$$\therefore \qquad \frac{|C|^2}{|A|^2} = \frac{4k^2}{(k + K)^2}$$

The current s is obtained from the formula

$$s = \frac{\hbar}{2mi}\left(\psi^* \frac{d\psi}{dx} - \psi \frac{d\psi^*}{dx}\right)$$

For the incident beam $\psi = A \exp(ikx)$, therefore

$$\text{incident current} = \frac{\hbar}{2mi}[A^* \exp(-ikx) \,.\, ikA \exp(ikx) - A \exp(ikx)\{-ikA^* \exp(-ikx)\}]$$

$$= \frac{\hbar k}{m}|A|^2$$

The transmitted beam is represented by $\psi = C \exp(iKx)$ and the transmitted current $= \hbar K/m|C|^2$. Therefore

$$\frac{\text{transmitted current}}{\text{incident current}} = \frac{K}{k}\frac{|C|^2}{|A|^2} = \frac{4kK}{(k+K)^2}$$

k and K are the momenta of the electrons before and after the potential step respectively, and, therefore, are directly proportional to the velocities. Therefore

$$\frac{\text{transmitted current}}{\text{incident current}} = \frac{4 \times 0{\cdot}25v \times v}{(1{\cdot}25v)^2}$$

$$\text{transmitted beam} = \underline{0{\cdot}64\ \mu\text{A}}$$

Problem 5.15

Derive an expression for the particle current density vector **S** defined by

$$\text{div}\,\mathbf{S} = -\frac{\partial}{\partial t}(\Psi\Psi^*)$$

where Ψ is a solution of Schrodinger's time-dependent equation

$$H\Psi = \left(-\frac{\hbar^2}{2m}\nabla^2 + V\right)\Psi = i\hbar\frac{d\Psi}{dt}$$

in which V is a scalar potential function.

A homogeneous beam of particles, each of energy E, is incident upon a one-dimensional rectangular potential barrier defined by

$$V(x) = 0 \text{ for } -\infty < x < 0$$
$$V(x) = V_0 \text{ for } 0 < x < l$$
$$V(x) = 0 \text{ for } l < x < +\infty$$

Show how to determine the ratio between the particle current densities in the transmitted and incident beams, and find the numerical value of this ratio when (*a*) $E = 4\,V_0/5$, (*b*) $E = 9\,V_0/5$, given that in case (*a*) the de Broglie wavelength of the incident wave is $2l$ and that the barrier has the same dimensions in each case.

(Take $\sinh \frac{1}{2}\pi = 2{\cdot}30$.)

[B.Sc. (Special), Pt. II, London, 1961]

SOLUTION

A diagram of the potential barrier is shown in Fig. 5.6. The Schrödinger equation is

$$\frac{d^2\psi}{dx^2} + \frac{2m}{\hbar^2}[E - V]\,\psi = 0$$

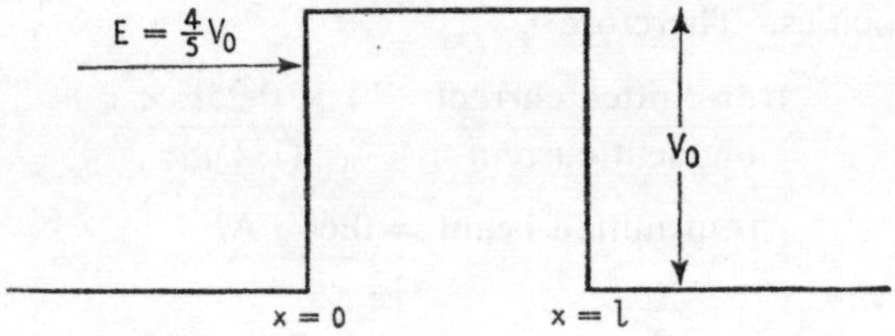

Fig. 5.6

For $x > l$, $V = 0$

$$\frac{d^2\psi}{dx^2} + k^2\psi = 0$$

where $k^2 = 2mE/\hbar^2$. Therefore

$$\psi = A \exp(ikx) \qquad (5.15.1)$$

The solution $\exp(-ikx)$ has no physical meaning.

For $0 < x < l$, $V = V_0$

$$\frac{d^2\psi}{dx^2} - K^2\psi = 0$$

where $K^2 = 2m(V_0 - E)/\hbar^2$. Therefore

$$\psi = B \exp(-Kx) + C \exp(Kx) \qquad (5.15.2)$$

The solution $B \exp(-Kx)$ is now acceptable since reflection can occur at the potential change at $x = l$.

For $x < 0$, $V = 0$

$$\psi = D \exp(ikx) + H \exp(-ikx) \qquad (5.15.3)$$

is an acceptable solution. Note that in this case the $-ve$ exponential must be included as there will be reflection from the potential change at $x = 0$.

Applying the boundary conditions at $x = 0$, Equations 5.15.2 and 5.15.3 give

$$D + H = B + C \quad (\psi \text{ continuous}) \qquad (5.15.4)$$

$$ik(D - H) = K(C - B) \quad \left(\frac{d\psi}{dx} \text{ continuous}\right) \qquad (5.15.5)$$

At $x = l$, Equations 5.15.1 and 5.15.2 give

$$A \exp(ikl) = B \exp(-Kl) + C \exp(Kl) \quad (\psi \text{ continuous}) \quad (5.15.6)$$

$$ikA \exp(ikl) = -K[B \exp(-Kl) - C \exp(Kl)] \quad \left(\frac{d\psi}{dx} \text{ continuous}\right) \qquad (5.15.7)$$

The ratio between the particle current densities in the transmitted and incident beams is T where

$$T = \frac{|A|^2}{|D|^2}$$

Using the above boundary conditions to eliminate H, B and C, Equations 5.15.4 and 5.15.5 give

$$D = \frac{B}{2}\left(1 + \frac{iK}{k}\right) + \frac{C}{2}\left(1 - \frac{iK}{k}\right) \qquad (5.15.8)$$

Adding Equations 5.15.6 and 5.15.7

$$C = \tfrac{1}{2} A \exp(ikl) \exp(-Kl) \left[1 + \frac{ik}{K}\right]$$

Subtracting Equation 5.15.7 from 5.15.6

$$B = \tfrac{1}{2} A \exp(ikl) \exp(Kl) \left(1 - \frac{ik}{K}\right)$$

Thus, substituting for B and C in Equation 5.15.8

$$D = \frac{A \exp(\mathrm{i}kl)}{4}\left[\exp(Kl)\left(1 - \frac{\mathrm{i}k}{K}\right)\left(1 + \frac{\mathrm{i}K}{k}\right) + \exp(-Kl)\left(1 - \frac{\mathrm{i}K}{k}\right)\left(1 + \frac{\mathrm{i}k}{K}\right)\right]$$

which can be arranged to give

$$D = \frac{A \exp(\mathrm{i}kl)}{4}\left[2\{\exp(Kl) + \exp(-Kl)\} - \mathrm{i}\left(\frac{k}{K} - \frac{K}{k}\right)\{\exp(Kl) - \exp(-Kl)\}\right]$$

$$\therefore \quad |D|^2 = \frac{|A|^2}{16}\left[4\{\exp(Kl) + \exp(-Kl)\}^2 + \left(\frac{k}{K} - \frac{K}{k}\right)^2\{\exp(Kl) - \exp(-Kl)\}^2\right]$$

$$= \frac{|A|^2}{16}\left\{16\cosh^2 Kl + 4\left(\frac{k}{K} - \frac{K}{k}\right)^2 \sinh^2 Kl\right\}$$

$$\therefore \quad T = \frac{|A|^2}{|D|^2} = \frac{1}{1 + \sinh^2 Kl + \frac{1}{4}\left(\frac{k}{K} - \frac{K}{k}\right)^2 \sinh^2 Kl}$$

$$= \frac{1}{\left[1 + \left\{1 + \frac{1}{4}\left(\frac{k}{K} - \frac{K}{k}\right)^2\right\}\sinh^2 kl\right]} \qquad (5.15.9)$$

(*a*) Using the information $E = \frac{4}{5}V_0$,

$$K = \sqrt{\left\{\frac{2m(V_0 - E)}{\hbar^2}\right\}} = \sqrt{\left(\frac{2mV_0}{5\hbar^2}\right)}$$

$$k = \sqrt{\left(\frac{2mE}{\hbar^2}\right)} = \sqrt{\left(\frac{8mV_0}{5\hbar^2}\right)}$$

$$\therefore \quad \frac{k}{K} = 2$$

The de Broglie wavelength, λ, of the incident electron beam is given by

$$\lambda = \frac{\mathbf{h}}{\sqrt{2mE}} = \frac{2\pi}{k} \quad \left(\text{since } \hbar = \frac{\mathbf{h}}{2\pi}\right)$$

In this case, $\lambda = 2l$, therefore

$$k = 2K = \frac{2\pi}{2l}$$

and
$$Kl = \frac{\pi}{2}$$

Substitution in Equation 5.15.9 gives

$$T = \frac{1}{\left\{1 + [1 + \frac{1}{4}(2 - \frac{1}{2})^2] \sinh^2 \frac{\pi}{2}\right\}}$$

$$= \frac{1}{1 + \frac{25}{16} \times (2{\cdot}3)^2} = \underline{0{\cdot}108}$$

(*b*) For case $E = \frac{9}{5}V_0$, i.e. $E > V_0$, it is left to the student to show that the transmission T is given by the formula

$$T = \frac{1}{1 + \left[1 + \frac{1}{4}\left(\frac{k}{K} + \frac{K}{k}\right)^2\right] \sin^2 Kl} \qquad (5.15.10)$$

In this case

$$K = \sqrt{\left\{\frac{2m(E - V_0)}{\hbar^2}\right\}} = \sqrt{\left(\frac{2m4V_0}{5\hbar^2}\right)}$$

$$k = \sqrt{\left(\frac{2mE}{\hbar^2}\right)} = \sqrt{\left(\frac{2m9V_0}{5\hbar^2}\right)}$$

$$\therefore \qquad \frac{K}{k} = \sqrt{\frac{4}{9}} = \frac{2}{3}$$

From part (*a*) of the problem

$$\lambda = \frac{2\pi\hbar}{\sqrt{(2m4/5V_0)}} = \frac{2\pi}{K} = 2l$$

Thus $Kl = \pi$ and $\sin Kl = \sin \pi = 0$. Substituting in Equation 5.15.10 gives transmission $\underline{T = 1}$.

Problem 5.16

Show that for a symmetrical potential distribution $V(x) = V(-x)$, the solutions of the one-dimensional Schrödinger equation must consist of even or odd functions, provided the energy levels are non-degenerate.

A particle of mass m is bound within a one-dimensional well of width $2a$ and depth V_0. Show that if $V_0a^2 = \hbar^2/2m$ there is only one bound energy level.

For a well having this value of V_0a^2, the position of the energy level is found to be $0{\cdot}57\ V_0$ above the bottom of the well. Show that the probability of locating the particle in the classically inaccessible region is approximately 32 per cent.

[B.Sc. (Special), Pt. II, London, 1960]

SOLUTION

The Schrödinger equation has the form

$$\frac{d^2\psi(x)}{dx} + \frac{2m}{\hbar^2}[E - V(x)]\psi(x) = 0$$

Substituting $x = -x$ and $V(x) = V(-x)$ we find both $\psi(x)$ and $\psi(-x)$ are acceptable solutions. If the energy levels are non-degenerate, then there can only be one linearly independent solution, i.e. the two solutions only differ by a constant. Therefore

$$\psi(x) = C\psi(-x)$$

Substituting $x = -x$,

$$\psi(-x) = C\psi(x) = C^2\psi(-x)$$

$$\therefore \qquad C = \pm 1$$

There are two types of solution

$\psi(x) = +\,\psi(-x)$ positive parity, even state
$\psi(x) = -\,\psi(-x)$ negative parity, odd state

Consider the case of the particle, mass m, bound in a potential well whose depth and width is $-V_0$ and $2a$ respectively (Fig. 5.7). For this particular case, $E < 0$ and $V = -V_0$ and the Schrödinger equation becomes

$$\frac{d^2\psi}{dx^2} + \frac{2m}{\hbar^2}[V_0 - E]\psi = 0$$

for $|x| < a$. The solution is of the form.

$$\psi(x) = A \sin kx + B \cos kx \tag{5.16.1}$$

where $k^2 = 2m(V_0 - E)/\hbar^2$.

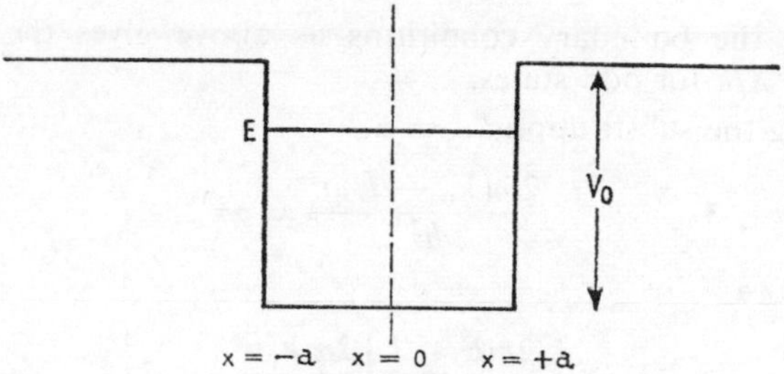

Fig. 5.7

For $|x| > a$, the Schrödinger equation is

$$\frac{d^2\psi}{dx^2} + \frac{2mE}{\hbar^2}\psi = 0$$

$$\therefore \qquad \psi(x) = C \exp(-K|x|) \tag{5.16.2}$$

where $K^2 = 2mE/\hbar^2$. The $+ve$ exponential is disregarded because it has no physical meaning. It has already been shown that if V is symmetric in x then the solutions may be classified as *even* and *odd*. Consider the *even* state solutions:

In this case $\psi(x) = \psi(-x)$, and so the general solution 5.16.1 for $|x| < a$ reduces to

$$\psi(x) = B \cos kx \tag{5.16.3}$$

Applying the boundary conditions at $x = a$, i.e. $\psi(x = a)$ and $\frac{d\psi}{dx}(x = a)$ must be continuous, then from Equations 5.16.2 and 5.16.3

$$C \exp(-Ka) = B \cos(ka)$$

$$-KC \exp(-Ka) = -kB \sin(ka)$$

Therefore $\tan ka = \frac{K}{k}$ for even states.

Consider the *odd* state solutions:
In this case

$$\psi(x) = -\psi(-x)$$

i.e. for $|x| < a$

$$\psi(x) = A \sin kx$$

Applying the boundary conditions as above gives the condition $\cot ka = K/k$ for odd states.

Making the substitution $ka = \alpha$,

$$\frac{2m(V_0 - E)a^2}{\hbar^2} = \alpha^2$$

Therefore

$$K = \frac{\sqrt{2mE}}{\hbar} = \frac{1}{a}\left[\frac{2mV_0a^2}{\hbar^2} - \alpha^2\right]^{\frac{1}{2}}$$

Under the conditions specified in the problem

$$\frac{2mV_0a^2}{\hbar^2} = 1$$

$\therefore$

$$Ka = (1 - \alpha^2)^{\frac{1}{2}}$$

and

$$\frac{K}{k} = \frac{(1 - \alpha^2)^{\frac{1}{2}}}{\alpha}$$

Thus the conditions for the solutions are now

$$\left.\begin{aligned} \tan\alpha &= \frac{(1-\alpha^2)^{\frac{1}{2}}}{\alpha} \\ \text{or} \quad \cot\alpha &= \frac{\alpha}{(1-\alpha^2)^{\frac{1}{2}}} \end{aligned}\right\} \text{even states}$$

and

$$\left.\begin{aligned} \cot\alpha &= -\frac{(1-\alpha^2)^{\frac{1}{2}}}{\alpha} \\ \text{or} \quad \tan\alpha &= -\frac{\alpha}{(1-\alpha^2)^{\frac{1}{2}}} \end{aligned}\right\} \text{odd states}$$

which can be solved graphically, see Figs. 5.8 and 5.9.

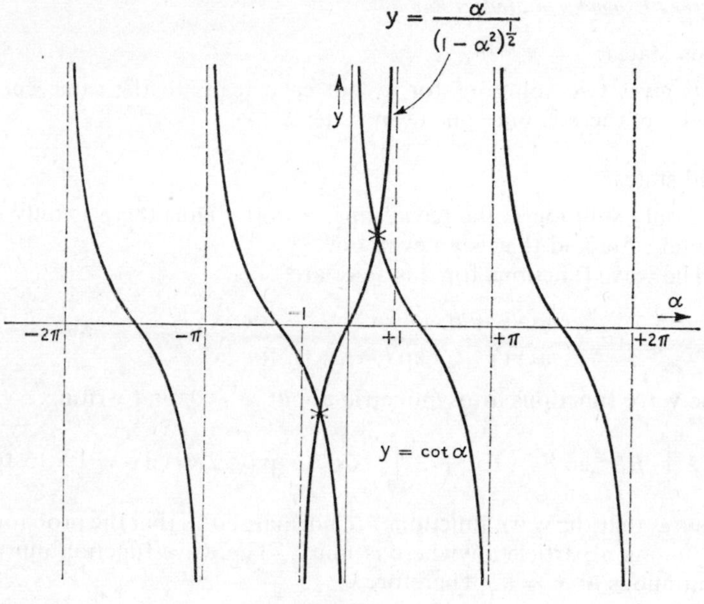

Fig. 5.8

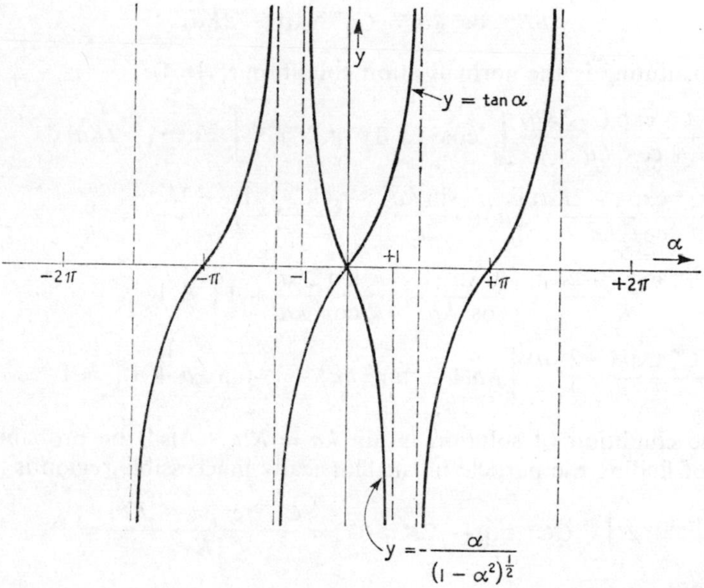

Fig. 5.9

Even states:

This gives two solutions for α, but each leads to the same energy level, i.e. there is only one even state.

Odd states:

The only solution is the trivial one, $\alpha = 0$. Thus there is only one bound state and that is an even one.

The wave functions for this state are

$$\psi(x) = B \cos kx \qquad \text{for } |x| < a$$
$$\psi(x) = C \exp(-K|x|) \quad \text{for } |x| > a$$

The wave functions are symmetric about $x = 0$, and writing

$$2\int_0^a BB^* \cos^2 kx \, dx + 2\int_a^\infty CC^* \exp(-2Kx) \, dx = 1 \qquad (5.16.4)$$

ensures that the wave functions are normalised so that the probability of finding a particle anywhere is unity. The wave function must be continuous at $x = a$. Therefore

$$B \cos ka = C \exp(-Ka)$$

and

$$B^* \cos ka = C^* \exp(-Ka)$$

$\therefore$

$$BB^* \cos^2 ka = CC^* \exp(-2Ka)$$

Substituting in the normalisation condition (5.16.4),

$$\frac{2CC^* \exp(-2Ka)}{\cos^2 ka}\int_0^a \cos^2 kx \, dx + 2CC^* \int_a^\infty \exp(-2Ka) \, dx = 1$$

$$\frac{CC^* \exp(-2Ka)}{\cos^2 ka}\left[a + \frac{\sin ka}{2k}\right] + \frac{CC^* \exp(-2Ka)}{K} = 1$$

$$\frac{CC^* \exp(-2Ka)}{K}\left[\frac{Ka}{\cos^2 ka} + \frac{K \sin 2ka}{k \cos^2 ka} + 1\right] = 1$$

$$\frac{CC^* \exp(-2Ka)}{K}\left[Ka(1 + \tan^2 ka) + \frac{K}{k}\tan ka + 1\right] = 1$$

The condition of solution is $\tan ka = K/k$. Also the probability P of finding the particle in the classically inaccessible region is

$$2\int_a^\infty CC^* \exp(-2Kx) \, dx = \frac{CC^* \exp(-2Ka)}{K} = P$$

Thus

$$P = \frac{1}{\left(1 + \frac{K^2}{k^2}\right)(1 + Ka)}$$

Writing P in terms of α,

$$\left(1 + \frac{K^2}{k^2}\right) = \frac{1}{\alpha^2} \quad \text{and} \quad Ka = (1 - \alpha^2)^{\frac{1}{2}}$$

$$\therefore \qquad P = \frac{\alpha^2}{1 + (1 - \alpha^2)^{\frac{1}{2}}}$$

Now

$$\alpha^2 = \frac{2m(V_0 - E)a^2}{\hbar^2} = \frac{2mV_0a^2}{\hbar^2} \times 0{\cdot}57$$

since the position of the bound level is $0{\cdot}57\ V_0$ above the bottom of the well.

Also $2mV_0a^2/\hbar^2 = 1$, so that $\alpha^2 = 0{\cdot}57$. Therefore

$$P = \frac{0{\cdot}57}{1 + (0{\cdot}43)^{\frac{1}{2}}} = 0{\cdot}34$$

i.e. $\underline{P = 34\%}$

Problem 5.17

State the boundary conditions which a wave function must satisfy at an abrupt change in potential.

Consider a potential of the form $V = V_0 > 0$ for $x > 0$ and $V = 0$ for $x \leqslant 0$. A beam of particles having energy E is incident from $-\infty$, travelling in the direction of $+ve\ x$. Calculate the fraction of the incident beam reflected and sketch a graph showing the variation of this fraction with E.

[B.Sc. Higher, Pt. II, III, Liverpool, 1963]

Ans. For the case $E > V_0$ the fraction reflected is $(1 - \alpha)^2/(1 + \alpha)^2$ where $\alpha = (1 - V_0/E)^{\frac{1}{2}}$. For $E < V_0$ the fraction reflected is unity.

Problem 5.18

A particle is constrained to move on a line on which its potential is everywhere zero except for a section of length a where it is V. Show that if V is negative there is at least one bound state having negative total energy, and sketch the form of the wavefunction for this state when a and V are small.

If the particle is an electron, and a and V are equal to 8×10^{-8} cm and 18 electron volts respectively, how many bound states are there?

[Nat. Science Tripos, Pt. I, Cambridge, 1958]

Ans. 6

OPERATOR FORMALISM

Problem 5.19

Show how the expectation value of a dynamical variable may be calculated when it is measured for a system in a state which is not an eigenstate of that variable.

A linear harmonic oscillator of mass m, subject to a restoring force bx, is in its lowest energy state, for which the wave function has the form

$$\psi_0(x) = A \exp\left(-\tfrac{1}{2}\alpha^2 x^2\right)$$

where $\alpha^2 = \sqrt{mb}/\hbar$.

Calculate the expectation values of x^2 and p^2 for this state, where p is the momentum. Hence show that the expectation values of the potential energy and the kinetic energy are each equal to $\frac{1}{4}\hbar\sqrt{(b/m)}$.

Comment on this result.

$$\left[\int_0^\infty \exp\left(-x^2\right) \mathrm{d}x = \sqrt{\pi}/2\right]$$

[B.Sc. (Special), Pt. II, London, 1959]

SOLUTION

The expectation value of x^2 is by definition

$$\langle x^2 \rangle = \frac{\int_{-\infty}^{+\infty} \psi_0^*(x) x^2 \psi_0(x)\, \mathrm{d}x}{\int_{-\infty}^{+\infty} \psi_0^*(x) \psi_0(x)\, \mathrm{d}x}$$

where in this case $\psi_0 = A \exp(-\frac{1}{2}\alpha^2 x^2)$. Thus

$$\langle x^2 \rangle = \frac{\int_{-\infty}^{+\infty} A^* \exp(-\frac{1}{2}\alpha^2x^2)x^2 A \exp(-\frac{1}{2}\alpha^2x^2)\, \mathrm{d}x}{\int_{-\infty}^{+\infty} A^* \exp(-\frac{1}{2}\alpha^2x^2) A \exp(-\frac{1}{2}\alpha^2x^2)\, \mathrm{d}x}$$

$$= \frac{A^*A \int_{-\infty}^{+\infty} \exp(-\alpha^2x^2)x^2\, \mathrm{d}x}{A^*A \int_{-\infty}^{+\infty} \exp(-\alpha^2x^2)\, dx}$$

As both integrals are even functions of x, then

$$\langle x^2 \rangle = \frac{\int_0^{\infty} \exp(-\alpha^2x^2)x^2\, \mathrm{d}x}{\int_0^{\infty} \exp(-\alpha^2x^2)\, \mathrm{d}x}$$

The value of the integral $\int_0^{\infty} \exp(-x^2)\, \mathrm{d}x = \sqrt{\pi}/2$ is given in the question. A straightforward substitution $x^2 = \alpha^2u^2$ leads to

$$\int_0^{\infty} \exp(-\alpha^2u^2)\, \mathrm{d}u = \frac{1}{2}\sqrt{\frac{\pi}{\alpha^2}}$$

Differentiating this integral with respect to α

$$\frac{\mathrm{d}}{\mathrm{d}\alpha}\int_0^{\infty} \exp(-\alpha^2u^2)\, \mathrm{d}u = \int_0^{\infty} -2\alpha u^2 \exp(-\alpha^2u^2)\, \mathrm{d}u$$

$$-\frac{1}{2\alpha}\frac{\mathrm{d}}{\mathrm{d}\alpha}\left(\frac{1}{2\alpha}\sqrt{\pi}\right) = \int_0^{\infty} u^2 \exp(-\alpha^2u^2)\, \mathrm{d}u$$

$$\int_0^{\infty} u^2 \exp(-\alpha^2u^2)\, \mathrm{d}u = \frac{1}{4\alpha^2}\sqrt{\frac{\pi}{\alpha^2}}$$

Therefore

$$\langle x^2 \rangle = \frac{\frac{1}{4\alpha^2}\sqrt{\left(\frac{\pi}{\alpha^2}\right)}}{\frac{1}{2}\sqrt{\left(\frac{\pi}{\alpha^2}\right)}} = \frac{1}{2\alpha^2}$$

$$\underline{\langle x^2 \rangle = \frac{\hbar}{2\sqrt{(mb)}}}$$

The expectation value for p is $\langle p^2 \rangle$. The operator for momentum is $(\hbar/\mathrm{i})\,(\mathrm{d}/\mathrm{d}x)$, therefore

$$\langle p^2 \rangle = \frac{\displaystyle\int_{-\infty}^{+\infty} A^* \exp(-\tfrac{1}{2}\alpha^2x^2)\left[-\hbar^2 \frac{\mathrm{d}^2}{\mathrm{d}x^2}\{A \exp(-\tfrac{1}{2}\alpha^2x^2)\}\right] \mathrm{d}x}{\displaystyle\int_{-\infty}^{+\infty} A^* \exp(-\tfrac{1}{2}\alpha^2x^2)\, A \exp(-\tfrac{1}{2}\alpha^2x^2)\, \mathrm{d}x}$$

Now

$$-\frac{\mathrm{d}^2}{\mathrm{d}x^2}[A \exp(-\tfrac{1}{2}\alpha^2x^2)] = -\frac{\mathrm{d}}{\mathrm{d}x}[-\alpha^2x^2 A \exp(-\tfrac{1}{2}\alpha^2x^2)]$$

$$= A \exp(-\tfrac{1}{2}\alpha^2x^2)\,[\alpha^2 - \alpha^4x^2]$$

Therefore

$$\langle p^2 \rangle = \frac{A^*A \displaystyle\int_{-\infty}^{+\infty} \exp(-\alpha^2x^2)[\alpha^2 - \alpha^4x^2]\, \mathrm{d}x}{A^*A \displaystyle\int_{-\infty}^{+\infty} \exp(-\alpha^2x^2)\, \mathrm{d}x}$$

Again the integrals are even functions of x, therefore

$$\langle p^2 \rangle = \frac{\hbar^2 \left[\displaystyle\int_0^{\infty} \alpha^2 \exp(-\alpha^2x^2)\, \mathrm{d}x - \int_0^{\infty} \alpha^4x^2 \exp(-\alpha^2x^2)\, \mathrm{d}x\right]}{\displaystyle\int_0^{\infty} \exp(-\alpha^2x^2)\, \mathrm{d}x}$$

All these integrals have been evaluated so that substituting the values obtained gives

$$\langle p^2 \rangle = \frac{\hbar^2 \left[\alpha^2 \dfrac{1}{2} \sqrt{\left(\dfrac{\pi}{\alpha^2}\right)} - \alpha^4 \dfrac{1}{4\alpha^2} \sqrt{\left(\dfrac{\pi}{\alpha^2}\right)}\right]}{\dfrac{1}{2}\sqrt{\left(\dfrac{\pi}{\alpha^2}\right)}} = \tfrac{1}{2}\,\hbar^2\alpha^2$$

Therefore

$$\underline{\langle p^2 \rangle = \tfrac{1}{2}\hbar\sqrt{(mb)}}$$

The mean potential energy

$$\tfrac{1}{2}\, b\, \langle x^2 \rangle = \frac{b}{2}\,\frac{\hbar}{2\sqrt{mb}} = \underline{\tfrac{1}{4}\,\hbar \sqrt{\frac{b}{m}}}$$

The mean kinetic energy

$$\frac{\langle p^2\rangle}{2m} = \frac{\hbar}{4m}\sqrt{mb} = \tfrac{1}{4}\hbar\sqrt{\frac{b}{m}}$$

i.e. the average P.E. = average K.E. = $\frac{1}{4}\hbar\sqrt{(b/m)}$.

Problem 5.20

Outline a method of obtaining the wave functions and energy levels of the hydrogen atom.

The ground state wave function of this atom is

$$(\pi a_0^3)^{-\frac{1}{2}} \exp\left(-\frac{r}{a_0}\right)$$

where a_0 is the radius of the Bohr orbit. Determine (*a*) the mean value, and (*b*) the most probable value of r in this state and comment on the significance of these results.

In what important respects do the wave functions of the 2*s* and 2*p* states of the hydrogen atom differ from that of the ground state?

[B.Sc. (Special), Pt. II, London, 1961]

SOLUTION

(*a*) The wave function

$$\psi = (\pi a_0^3)^{-\frac{1}{2}} \exp\left(-\frac{r}{a_0}\right)$$

is normalised. Therefore the mean value of r (i.e. $\bar{r}$) is given by the formula

$$\bar{r} = \int \psi^* r \psi \, \mathrm{d}V$$

The integration must be carried out over all space. In spherical co-ordinates $\mathrm{d}V = r^2 \sin\theta \, \mathrm{d}\theta \, \mathrm{d}\phi \, \mathrm{d}r$

$$\therefore \qquad \bar{r} = \frac{1}{(\pi a_0^3)} \int_0^\infty \exp\left(-\frac{2r}{a_0}\right) r^3 \, \mathrm{d}r \int_0^\pi \sin\theta \, \mathrm{d}\theta \int_0^{2\pi} \mathrm{d}\phi$$

$$= \frac{4}{a_0^3} \int_0^\infty r^3 \exp\left(-\frac{2r}{a_0}\right) \mathrm{d}r$$

The integral

$$\int_0^\infty r^3 \exp\left(-\frac{2r}{a_0}\right) dr = \left(\frac{a_0}{2}\right)^4 3!$$

$$\therefore \qquad \bar{r} = \frac{4}{a_0^3}\left(\frac{a_0}{2}\right)^4 3! = \tfrac{3}{2}\, a_0$$

i.e. the mean value of r is $\frac{3}{2}a_0$ where a_0 is the radius of the Bohr orbit.

(*b*) The probability that the electron lies between a radius r and $r + dr$ is $|\psi(r)|^2 4\pi r^2\, dr$. The probability density function

$$p(r) = 4\pi r^2 |\psi(r)|^2$$

$$= \frac{4\pi r^2}{\pi a_0^3} \exp\left(-\frac{2r}{a_0}\right) = \frac{4r^2}{a_0^3} \exp\left(-\frac{2r}{a_0}\right)$$

The most probable value of r occurs for the maximum value of $p(r)$. Therefore

$$\frac{dp(r)}{dr} = \frac{d}{dr}\left[\frac{4r^2}{a_0^3} \exp\left(-\frac{2r}{a_0}\right)\right] = 0$$

gives the most probable value of r.

$$\frac{4}{a_0^3}\left\{2r \exp\left(-\frac{2r}{a_0}\right) + r^2\left(-\frac{2}{a_0}\right) \exp\left(-\frac{2r}{a_0}\right)\right\} = 0$$

$$2ra_0 = 2r^2$$

$$\underline{r = a_0}$$

Therefore the most probable value of r is the Bohr orbit.

Problem 5.21

Show that acceptable solutions of Schrödinger's equation for the linear harmonic oscillator can only be obtained if the energy is restricted to the values

$$E_n = (n + \tfrac{1}{2})\mathbf{h}\nu \quad (n = 0, 1, 2, \ldots)$$

where ν is the classical frequency.

If a particle moves in three dimensions in a potential field given by $V = \frac{1}{2}kr^2$ where r is the distance from a fixed point and k is a constant, determine the energy levels and their degree of degeneracy for this isotropic harmonic oscillator.

[B.Sc. (Special), Pt. II, London, 1961]

Ans. Working in cartesian co-ordinates it may be shown that $E = (n_x + n_y + n_z + 3/2)\hbar\sqrt{(k/m)} = (N + 3/2)\hbar\sqrt{(k/m)}$. The degree of degeneracy is equal to the number of ways n_x, n_y, n_z can be added to give N. The integers n_x, n_y, n_z must be non-negative. E.g. $N = 0$, non-degenerate; $N = 1$, three degenerate levels; $N = 2$, six degenerate levels.

6
Special Theory of Relativity

MICHELSON-MORLEY EXPERIMENT

Problem 6.1

Describe a form of the Michelson and Morley experiment.

Calculate the maximum displacement in the fringe system which might be expected on the assumption that light waves are propagated with a velocity 3×10^5 km sec^{-1} in a medium which is at rest relative to the sun, using an apparatus in which the interfering beams are each 10 metres long. The distance from the earth to the sun may be taken as $1{\cdot}5 \times 10^8$ km.

How do the postulates of the special theory of relativity account for the fact that no fringe shift is observed?

[B.Sc. (Special), Pt. II, London, 1958]

SOLUTION

The apparatus of Michelson and Morley consists of two light paths 1 and 2 which are perpendicular to each other (Fig. 6.1).

The incident light beam is split into two by means of a half silvered mirror along these two paths. The apparatus is arranged so that one arm (in this case arm 2) is perpendicular to the earth's surface.

Let the time taken for the light to travel along path 1 and back be t_1.

$$t_1 = \frac{l}{\mathbf{c} - v} + \frac{l}{\mathbf{c} + v}$$

$$= \frac{2l\mathbf{c}}{(\mathbf{c} + v)(\mathbf{c} - v)} = \frac{2l}{\mathbf{c}}\left[\frac{1}{1 - \frac{v^2}{\mathbf{c}^2}}\right]$$

The time taken to travel along path 2 and back is t_2 and the velocity of light relative to arm 2 is $\sqrt{(c^2 - v^2)}$. Therefore

$$t_2 = \frac{2l}{\sqrt{(c^2 - v^2)}} = \frac{2l}{c}\left[\frac{1}{\sqrt{\left(1 - \frac{v^2}{c^2}\right)}}\right]$$

Consider v, the velocity of the earth: if r is the radius of the earth's orbit and T is the time of one period then

$$v = \frac{2\pi r}{T}$$

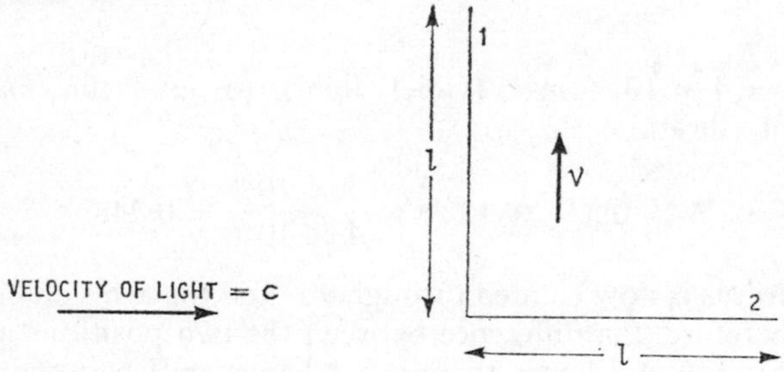

Fig. 6.1

Substituting the values $T = 365 \times 8{\cdot}64 \times 10^4$ sec and $r = 1{\cdot}5 \times 10^8$ km

$$v = \frac{2 \times 1{\cdot}5 \times 10^8}{365 \times 8{\cdot}64 \times 10^4} = 29{\cdot}87 \text{ km/sec}$$

Now $v \ll c$ and, therefore, the expressions for the times t_1 and t_2 can be expressed in series form

$$t_1 = \frac{2l}{c}\left[1 + \frac{v^2}{c^2} + \ldots\right]$$

$$t_2 = \frac{2l}{c}\left[1 + \frac{v^2}{2c^2} + \ldots\right]$$

and terms of higher order than v^2 neglected. Therefore the time interval

$$\Delta t = t_1 - t_2 = \frac{2l}{c}\left[\frac{v^2}{c^2} - \frac{v^2}{2c^2}\right] = \frac{l}{c}\frac{v^2}{c^2}$$

Substituting the numerical values $l = 10^{-2}$ km, $\mathbf{c} = 3 \times 10^5$ km/sec, $v = 29{\cdot}87$ km/sec, then

$$\Delta t = \frac{10^{-2}}{3 \times 10^5}\left(\frac{29{\cdot}87}{3 \times 10^5}\right)^2 = 0{\cdot}331 \times 10^{-15} \text{ sec}$$

The number of vibrations in this time interval is

$$N = \frac{\Delta t}{T} = \nu\,\Delta t$$

where ν is the frequency of the light. Therefore

$$N = \frac{\mathbf{c}}{\lambda}\,\Delta t$$

Using $\lambda = 4 \times 10^{-5}\,\text{cm}^{-1}$ (violet light) to give the maximum number of vibrations

$$N = 0{\cdot}331 \times 10^{-15} \times \frac{3 \times 10^{10}}{4 \times 10^{-5}} = 0{\cdot}248$$

The apparatus is now rotated through 90° and the arms change their roles. Therefore, the difference between the two positions amounts to $\frac{1}{2}$ a vibration and hence there is a $\frac{1}{2}$ fringe shift between the two positions.

THE LORENTZ TRANSFORMATION

Problem 6.2

State what is meant by a Galilean transformation, and a Lorentz transformation, and explain carefully the term Lorentz contraction.

A radioactive nucleus decays into two equal fragments each of which has a velocity 0·6 **c** with respect to the centre of mass. What will be the velocity of each fragment as seen by a stationary observer if the original nucleus is travelling at 0·5 **c** and the fragments separate along the line of motion?

[B.Sc. Inter., Manchester, 1963]

SOLUTION

S is the inertial frame at rest with respect to the observer and S' the inertial frame at rest with respect to the centre of mass (Fig. 6.2).

Let the nucleus decay into two equal parts, A and B, which have equal but opposite velocities u and take the time $t' = 0$ as the time at which the nucleus decays in the frame S'.

Reference frame S':

The locus of event A is represented by the four-vector $(t', -ut', 0, 0)$ and B by $(t', ut', 0, 0)$.

Reference Frame S:

The locus of event A is represented by the four-vector $(t, u_A t, 0, 0)$ and B by $(t, u_B t, 0, 0)$, where u_A and u_B are the velocities of A and B

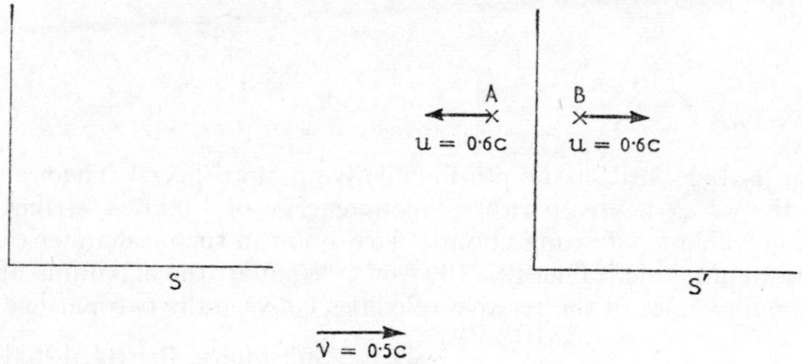

Fig. 6.2

respectively. The transformation between the two frames of reference is given by

$$x = \gamma(x' + vt')$$

$$t = \gamma\left(t' + \frac{x'v}{c^2}\right)$$

where $\gamma = 1/\sqrt{(1 - v^2/c^2)}$.

For particle A, $x' = -ut'$, therefore

$$x = \gamma t'(-u + v)$$

and

$$t = \gamma t'\left(1 - \frac{uv}{c^2}\right)$$

The velocity of A

$$u_A = \frac{x}{t} = \frac{v - u}{1 - \dfrac{uv}{c^2}}$$

For particle B, $x' = +ut'$. Therefore the velocity of B

$$u_B = \frac{v+u}{1+\dfrac{uv}{\mathbf{c}^2}}$$

Substituting the values $v = 0{\cdot}5\ \mathbf{c}$, $u = 0{\cdot}6\ \mathbf{c}$,

$$u_A = -\frac{0{\cdot}1}{0{\cdot}7}\mathbf{c} \quad \text{and} \quad u_B = \frac{1{\cdot}1}{1{\cdot}3}\mathbf{c}$$

Therefore, particles A and B appear to move in opposite directions with velocities $\underline{0{\cdot}143\ \mathbf{c}}$ and $\underline{0{\cdot}846\ \mathbf{c}}$ respectively.

Problem 6.3

What is the basic assumption underlying the Special Theory of Relativity? A neutron with a kinetic energy of 1·00 BeV strikes a proton which is vibrating about a fixed point in such a manner that its maximum kinetic energy is 100 MeV. What are the maximum and minimum values of the relative velocities between the two particles?

[B.Sc. Subsidiary, Bristol, 1962]

SOLUTION

The kinetic energy of a particle of rest mass, m_0, is

$$m_0\mathbf{c}^2\left[\frac{1}{\left(1-\dfrac{v^2}{\mathbf{c}^2}\right)^{\frac{1}{2}}}-1\right]$$

where v is the velocity of the particle.

Assuming for the purpose of this problem that the rest mass of the proton and neutron are equal and that the magnitude is 940 MeV, then

$$\text{K.E. of neutron} = 1{\cdot}0\times10^3\ \text{MeV} = 940\left[\frac{1}{\left(1-\dfrac{v_n^2}{\mathbf{c}^2}\right)^{\frac{1}{2}}}-1\right]$$

$$\left(1-\frac{v_n^2}{\mathbf{c}^2}\right)^{-\frac{1}{2}}-1 = \frac{10^3}{940}$$

$$\frac{v_n}{c} = \left(\frac{3{\cdot}26}{4{\cdot}26}\right)$$

$$\therefore \qquad v_n = 0{\cdot}875\ \mathbf{c}$$

Similarly

$$\text{K.E. of proton} = 100\ \text{MeV} = m_0\mathbf{c}^2\left[\frac{1}{\left(1 - \frac{v_p^2}{\mathbf{c}^2}\right)^{+\frac{1}{2}}} - 1\right]$$

$$\left(1 - \frac{v_p^2}{\mathbf{c}^2}\right)^{-\frac{1}{2}} - 1 = \frac{10^2}{940}$$

$$\therefore \qquad \frac{v_p}{\mathbf{c}} = \left(\frac{0{\cdot}224}{1{\cdot}224}\right)^{\frac{1}{2}}$$

$$v_p = 0{\cdot}428\ \mathbf{c}$$

The normal rule for the composition of velocities (see Problem 6.2) gives

$$\text{Max. relative velocity} = \frac{v_p + v_n}{1 + \frac{v_p v_n}{\mathbf{c}^2}}$$

$$= \frac{(0{\cdot}428 + 0{\cdot}875)\mathbf{c}}{1 + 0{\cdot}428 \times 0{\cdot}875} = \underline{0{\cdot}948\ \mathbf{c}}$$

$$\text{Min. relative velocity} = \frac{v_n - v_p}{1 - \frac{v_p v_n}{\mathbf{c}^2}}$$

$$= \frac{(0{\cdot}875 - 0{\cdot}428)\mathbf{c}}{1 - (0{\cdot}428 \times 0{\cdot}875)} = \underline{0{\cdot}715\ \mathbf{c}}$$

Problem 6.4

In a particular frame of reference S, two particles A and B are seen to have velocities V_A and V_B, not necessarily in the same direction.

Assuming the Lorentz transformation between the frames of reference, deduce the velocity of B in a frame of reference S' in which particle A is stationary.

If in frame S the angle between V_A and V_B is 90° and $V_A = V_B = \frac{1}{2}\mathbf{c}$, what is the velocity of B and the angle between the motions in the frame S'?

[B.Sc. Higher, Pt. II, III, Liverpool, 1963]

SOLUTION

Choose a frame of reference S (Fig. 6.3) such that the velocity V_A is

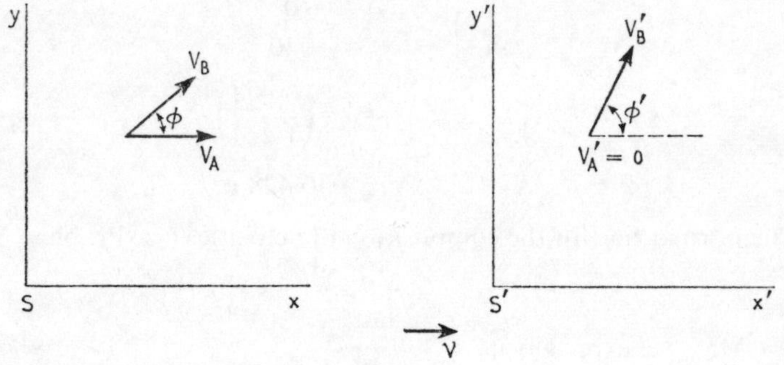

Fig. 6.3

directed along the x-axis. In this frame let B move with a velocity V_B at an angle ϕ to the x-axis.

In the new frame of reference S', the particle A has zero velocity, and B has a velocity V'_B at an angle ϕ' to the x'-axis. Let S' move with a velocity v with respect to S.

Particle A

In the frame S the locus of A is represented by the 4-vector $(t, u_a t, y, z)$ and in S' by $(t', u'_a t', y, z)$. Note that the y and z co-ordinates remain unchanged.

The transformation between S and S' is given by the Lorentz formulae

$$x' = \gamma(x - vt) \tag{6.4.1}$$

$$t' = \gamma\left(t - \frac{xv}{\mathbf{c}^2}\right)^{\frac{1}{2}} \tag{6.4.2}$$

where $\gamma = 1/\sqrt{(1 - v^2/c^2)}$. Substituting the coordinate $x = u_a t$ in Equations 6.4.1 and 6.4.2

$$x' = \gamma t(u_a - v)$$

$$t' = \gamma t\left(1 - \frac{u_a v}{c^2}\right) \tag{6.4.3}$$

The velocity

$$u'_a = \frac{x'}{t'} = \frac{u_a - v}{1 - \dfrac{u_a v}{c^2}} \tag{6.4.4}$$

Therefore for u'_a to be zero, v must equal u_a.

Particle B

In frame S the locus of B is represented by the vector $(t, u_b t, v_b t, z)$ and in S' by $(t', u_b t', v_b t', z)$ where u_b and v_b refer to the x and y components of the velocity V_B.

From the previous part of the problem (Equation 6.4.4) it can be inferred that

$$u_b = \frac{u_b - v}{1 - \dfrac{u_b v}{c^2}} = \frac{u_b - u_a}{1 - \dfrac{u_b u_a}{c^2}}$$

since $v = u_a$. It is noted that the y-coordinate must be the same in both reference frames. Therefore

$$v_b t = v'_b t'$$

$$v'_b = v_b\left(\frac{t}{t'}\right) \tag{6.4.5}$$

Substituting for t/t' in Equation 6·4·5 from Equation 6.4.3 with $v = u_a$,

$$v'_b = \frac{v_b}{\gamma\left(1 - \dfrac{u_b u_a}{c^2}\right)}$$

Therefore

$$V'^2_B = v'^2_b + u'^2_b$$

$$= \frac{1}{\left(1 - \dfrac{u_a v_a}{c^2}\right)^2}\left[\frac{v_b^2}{\gamma^2} + (u_b - u_a)^2\right] \tag{6.4.6}$$

The angle ϕ' is given by

$$\tan\phi' = \frac{v_b'}{u_b'} = \frac{v_b}{\gamma(u_b - u_a)}$$

For the special case quoted in the question, i.e. $\phi = 90°$, $V_A = V_B = \frac{1}{2}\mathrm{c}$,

$$u_a = \tfrac{1}{2}\mathrm{c},\ u_b = 0,\ v_a = 0,\ v_b = \tfrac{1}{2}\mathrm{c}$$

Now

$$\gamma = \frac{1}{\sqrt{\left(1 - \frac{v^2}{\mathrm{c}^2}\right)}} = \frac{1}{\sqrt{\left(1 - \frac{u_a^2}{\mathrm{c}^2}\right)}} = \frac{2}{\sqrt{3}}$$

Substituting these values into Equation 6.4.6 gives

$$V_B'^2 = \left[\frac{\frac{1}{4}\mathrm{c}^2}{4}3 + \frac{\mathrm{c}^2}{4}\right]$$

$$\therefore \qquad \underline{V_B' = \frac{\sqrt{7}\mathrm{c}}{4}}$$

Similarly

$$\tan\phi' = \frac{\frac{1}{2}\mathrm{c}\sqrt{3}}{2(-\frac{1}{2}\mathrm{c})} = -\frac{\sqrt{3}}{2}$$

$$\underline{\phi' = 150°}$$

i.e. V_B' is in a direction 150° to the +ve x'-axis.

Problem 6.5

Assuming the Lorentz transformation obtain formulae for the addition of velocities.

Experiments in nuclear physics are usually performed by bombarding a particle at rest with another particle from an accelerator. In analysing experiments, however, it is most physical to use the frame in which the centre of mass is at rest.

If the two particles involved have equal masses and undergo elastic scattering show that the velocity u and the scattering angle θ of the bombarding particle in the centre of mass frame are related to the corresponding quantities ω, Θ in the laboratory frame by:

$$\beta(u) = \left[\frac{1 + \beta(\omega)}{2}\right]^{\frac{1}{2}}$$

and $$\tan\left(\frac{\theta}{2}\right) = \beta(u)\tan\Theta$$

where $\beta(v) = 1/\sqrt{(1 - v^2/c^2)}$.

[B.Sc. Higher, Pt. II, III, Liverpool, 1963]

SOLUTION

The collision process is represented diagrammatically in Fig. 6.4 in

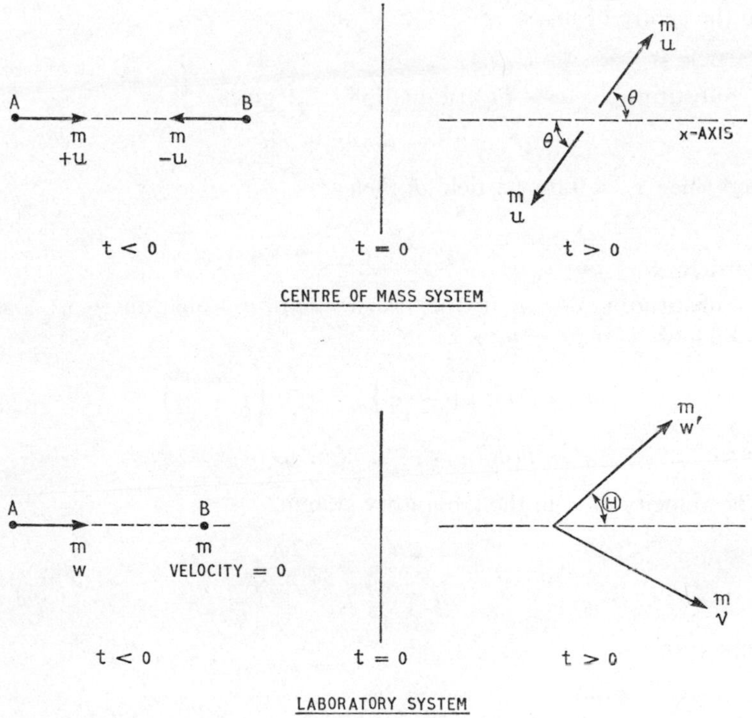

Fig. 6.4

both centre of mass and laboratory frames of reference with the collision assumed to take place at a time $t = 0$.

Before collision ($t < 0$):

In the centre of mass system the locus of A is represented by the four-vector $(t, ut, 0, 0)$ and B by $(t, -ut, 0, 0)$. The respective vectors in the laboratory system are $(t', \omega t', 0, 0)$ for A and $(t', 0, 0, 0)$ for B.

The Lorentz transformation gives the relationship between the two frames of reference

$$x' = \beta(v)[x - vt] \tag{6.5.1}$$

$$t' = \beta(v)\left[t - \frac{xv}{c^2}\right] \tag{6.5.2}$$

where the laboratory system is moving with a velocity v with respect to the centre of mass.

Particle B

Substituting $x = -ut$ in Equation 6.5.1 gives

$$x' = -\beta(v)[u + v]t$$

and since $x' = 0$ for particle B, then

$$v = -u$$

Particle A

Substituting $x = ut$ in the transformation Equations 6.5.1 and 6.5.2 and also $v = -u$,

$$t' = \beta(v)t\left(1 - \frac{uv}{c^2}\right) = \beta(v)t\left(1 + \frac{u^2}{c^2}\right)$$

$$x' = \beta(v)t(u - v) = \beta(v)t\,2u$$

The velocity of A in the laboratory system

$$\omega = \frac{x'}{t'} = \frac{2u}{1 + \dfrac{u^2}{c^2}} = \frac{2uc}{c^2 + u^2}$$

therefore

$$\beta(\omega) = \frac{1}{\sqrt{\left(1 - \dfrac{\omega^2}{c^2}\right)}} = \frac{1}{\sqrt{\left(1 - \dfrac{4u^2c^2}{(c^2 + u^2)^2}\right)}}$$

$$= \frac{c^2 + u^2}{c^2 - u^2} = \frac{2c^2}{c^2 - u^2} - 1$$

$$\therefore \qquad \frac{1 + \beta(\omega)}{2} = \frac{1}{1 - \dfrac{u^2}{c^2}} = \beta^2(u)$$

$$\therefore \qquad \beta(u) = \sqrt{\left\{\frac{1 + \beta(\omega)}{2}\right\}}$$

After Collision ($t > 0$):

Considering particle A in the centre of mass system, the locus of A is represented by $(t, ut \cos \theta, ut \sin \theta, 0)$ whereas in the laboratory system it is represented by $(t', \omega' t' \cos \Theta, \omega' t' \sin \Theta, 0)$. (The velocity of A has been resolved into its components in both systems.)

Substituting the values $x = ut \cos \theta$ and also $v = -u$ into the transformation Equations 6.5.1 and 6.5.2

$$x' = \beta(u) \,.\, ut(1 + \cos \theta) \tag{6.5.3}$$

$$t' = \beta(u) \,.\, t\left(1 + \frac{u^2 \cos \theta}{\mathbf{c}^2}\right) \tag{6.5.4}$$

Then
$$\frac{x'}{t'} = \omega' \cos \Theta = \frac{u(1 + \cos \theta)}{\left(1 + \frac{u^2}{\mathbf{c}^2} \cos \theta\right)} \tag{6.5.5}$$

Since there is no relative motion between the two frames of reference along the y-axis, the y co-ordinates must be equal.

$$ut \sin \theta = \omega' t' \sin \Theta$$

or
$$\omega' \sin \Theta = u \sin \theta \left(\frac{t}{t'}\right)$$

Substituting for t/t' from Equation 6.5.4

$$\omega' \sin \Theta = \frac{u \sin \theta}{\beta(u)\left[1 + \frac{u^2}{\mathbf{c}^2} \cos \theta\right]} \tag{6.5.6}$$

Dividing Equation 6.5.6 by 6.5.5

$$\beta(u) \tan \Theta = \frac{\sin \theta}{1 + \cos \theta} = \frac{2 \sin \frac{\theta}{2} \cos \frac{\theta}{2}}{2 \cos^2 \frac{\theta}{2}}$$

Therefore
$$\underline{\beta(u) \tan \Theta = \tan \frac{\theta}{2}}$$

Problem 6.6

The parameter **c** occurs in relativity transformations. What is the significance of this parameter? How may it be determined without reference to the velocity of electromagnetic waves?

Two space ships are travelling nose-first on parallel but oppositely directed paths with a relative velocity of 0·98**c**. The name of one space ship is printed on its side in letters 20 cm wide and 20 cm high. The telegraphist on the other space ship transmits its name in morse at 20 words per minute. Determine the shape and size of the letters and the speed of the morse as observed from the opposing vehicles as they pass.

[B.Sc. (Honours), Pt. II, Manchester, 1963]

SOLUTION

Let S and S' be the respective frames of reference for the two space ships (Fig. 6.5). The x co-ordinates of the letters on the space ship

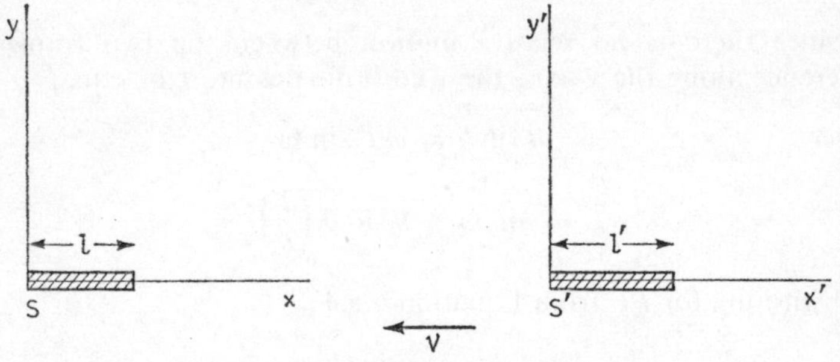

Fig. 6.5

S' are $(0,\ l')$. The Lorentz transformation between the frames S' and S is

$$t' = \gamma\left(t + \frac{xv}{\mathbf{c}^2}\right) \tag{6.5.1}$$

$$x' = \gamma(x + vt) \tag{6.5.2}$$

The observation of observer S is carried out at a time $t = t' = 0$, i.e. when the reference frames are coincident. Equation 6.5.2 immediately gives

$$x' = \gamma x$$

Therefore, the spaceship appears to have a length $x = l = x'/\gamma$, i.e. $l = l'\sqrt{(1 - v^2/c^2)}$.

Substituting the values $l' = 20$ cm, $v = 0{\cdot}98c$

$$l = 20\sqrt{\{1 - (0{\cdot}98)^2\}} = 3{\cdot}98 \text{ cm}$$

The height of the letters remains unaltered because there is no relative motion in this direction. Thus the letters appear to be 20 cm high and 4 cm wide according to an observer in spaceship S.

Consider the rate at which observer S' will receive the morse signals. In a small time interval Δt, observer S will see the spaceship S' move a distance $x = -v\Delta t$. Equation 6.5.1 gives

$$\Delta t' = \gamma\left(\Delta t + \frac{xv}{c^2}\right)$$

$$= \gamma\Delta t\left(1 - \frac{v^2}{c^2}\right) = \frac{\Delta t}{\gamma}$$

If f and f' are the frequencies of signals sent from ship S and S' respectively, then since the frequencies are inversely proportional to the time intervals

$$f' = \gamma f = \frac{20}{\sqrt{\{1 - (0{\cdot}98)^2\}}} = \underline{100 \text{ words/min}}$$

Problem 6.7

Derive the relation, given by Einstein's theory of special relativity, between the coordinates and time of a given event, as seen in two inertial frames of reference which are in uniform relative motion with velocity v in the positive x-direction. Show how this relation leads to the concept of 'time dilation'.

A neutral particle of mass 200 m_e is produced by cosmic rays at a height in the atmosphere of 10 km. If its lifetime at rest is 2×10^{-6} sec deduce the minimum energy in electron volts which the particle must have to survive to sea level.

[B.Sc. (Honours), Pt. I, Durham, 1963]

SOLUTION

Let S and S' be the inertial frame of reference for the earth and particle respectively (Fig. 6.6). The transformation between the two frames is given by equations

$$t' = \gamma\left(t + \frac{xv}{c^2}\right) \tag{6.7.1}$$

$$x' = \gamma(x + vt) \tag{6.7.2}$$

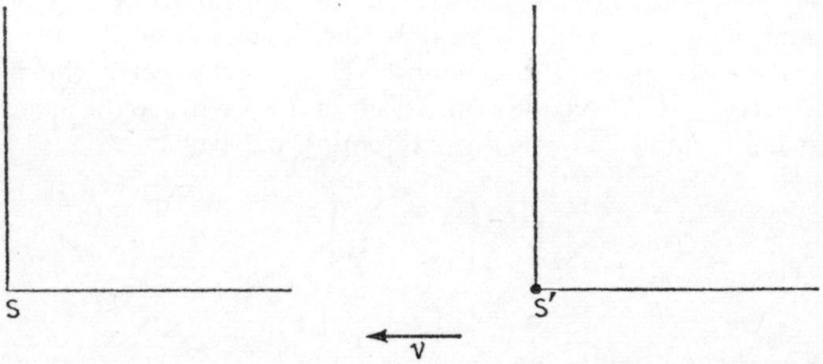

Fig. 6.6

The particle arrives at the earth's surface at $t = 0$. For $t < 0$

$$x' = 0 = \gamma(x + vt)$$

therefore

$$x = -vt$$

Substituting this value of x into Equation 6.7.1 gives

$$t' = \gamma t\left(1 - \frac{v^2}{c^2}\right) = \frac{t}{\gamma}$$

Thus the time intervals Δt and $\Delta t'$ relating to the inertial frames S and S' respectively are connected by the formula

$$\Delta t' = \frac{\Delta t}{\gamma}$$

Now $\Delta t'$ the mean lifetime of a particle at rest equals 2×10^{-6} sec. Therefore

$$\Delta t = \gamma \times 2 \times 10^{-6}\ \text{sec}$$

In this time interval the particle must travel a distance of 10 km ($= 10^6$ cm), i.e.

$$v = \frac{10^6}{\gamma(2 \times 10^{-6})}$$

$$\therefore \qquad \gamma v = 0{\cdot}5 \times 10^{12}$$

Substituting $\mathbf{c} = 3 \times 10^{10}$ cm/sec, squaring and rearranging gives

$$v^2 = \frac{10^{24}}{4\left(1 + \frac{10^4}{36}\right)}$$

or
$$1 - \frac{v^2}{\mathbf{c}^2} = 1 - \frac{10^4}{36 + 10^4} = \frac{36}{36 + 10^4}$$

$$\therefore \qquad \sqrt{\left(1 - \frac{v^2}{\mathbf{c}^2}\right)} \approx \frac{6}{10^2}$$

The K.E. of the particle $= \mathbf{mc}^2 \left[\frac{1}{\sqrt{\left(1 - \frac{v^2}{\mathbf{c}^2}\right)}} - 1\right]$

where $\mathbf{m}$, the rest mass of the particle, is 200 m_e, i.e.

$$\begin{aligned} \text{rest mass energy} &= 200 \times m_e\mathbf{c}^2 \\ &= 200 \times 0{\cdot}511 \text{ MeV} \\ &= 102{\cdot}2 \text{ MeV} \end{aligned}$$

$$\begin{aligned} \therefore \qquad \text{K.E. of particle} &= 102{\cdot}2 \left[\frac{10^2}{6} - 1\right] \\ &= 1{,}600 \text{ MeV} \\ &= 1{\cdot}6 \text{ GeV} \end{aligned}$$

$\therefore$ The K.E. of the particle must equal at least 1·6 GeV.

Problem 6.8

Describe the Doppler effect, and derive the relativistic formula for the longitudinal Doppler effect.

A space probe travelling directly away from the earth contains a transmitter radiating at a constant frequency of 10^9 c/s. The frequency of the signals reaching the earth can be measured with an accuracy of ± 1 c/s. At what speed relative to the earth will the relativistic Doppler shift differ measurably from that expected classically?

[B.Sc. (Honours), Manchester, 1963]

SOLUTION

The frequency of the signals from the probe according to the classical theory would be ν_c, where

$$\nu_c = \nu\left(1 - \frac{v}{\mathbf{c}}\right)$$

v being the velocity of the source. Taking into account relativistic effects the formula becomes

$$\nu_R = \nu\left(1 - \frac{v}{\mathbf{c}}\right)\gamma$$

where γ has its usual significance (see Problem 6.2). Thus

$$\frac{1}{\gamma^2} = 1 - \frac{v^2}{\mathbf{c}^2} = \left(\frac{\nu_c}{\nu_R}\right)^2$$

Therefore

$$\frac{v^2}{\mathbf{c}^2} = \frac{\nu_R^2 - \nu_c^2}{\nu_R^2} = \frac{(\nu_R + \nu_c)(\nu_R - \nu_c)}{\nu_R^2}$$

In this case $\nu_R \approx \nu_c$ and $\nu_R - \nu_c = \Delta\nu$. Therefore

$$\frac{v^2}{\mathbf{c}^2} = \frac{2\Delta\nu}{\nu_c} = \frac{2\Delta\nu}{\nu\left(1 - \frac{v}{\mathbf{c}}\right)}$$

or

$$-\frac{v^3}{\mathbf{c}^3} + \frac{v^2}{\mathbf{c}^2} = \frac{2\Delta\nu}{\nu}$$

The detector will only resolve frequency sources differing by 2 c/s, i.e. $\Delta\nu = 2$ c/s. Also assuming $v/\mathbf{c} \ll 1$, then the cubic term may be neglected. Therefore

$$\frac{v^2}{\mathbf{c}^2} = \frac{2 \times 2}{10^9}$$

$$\frac{v}{\mathbf{c}} = \sqrt{40} \times 10^{-5}$$

$$v = 3\sqrt{40} \times 10^5 = \underline{1{\cdot}9 \times 10^6 \text{ cm/sec}}$$

Problem 6.9

Derive the Lorentz transformations by means of the principle of relativity.

An observer A at rest on the ground sees two others, B and C, just above him, travelling in opposite directions with velocities v_1 and v_2 respectively. With what velocity does C travel with respect to B?

In a particular case, when B moves with respect to A with velocity $\sqrt{3}\mathbf{c}/2$ and C with velocity $\mathbf{c}/2$, all the observers set their watches at a zero hour when B and C are over A. When B observes that an interval of 10 sec has elapsed what is the interval according to A and C?

How far has B moved according to these observers and what is the distance between A and B according to C at the end of the interval?

[B.Sc. (Special) Pt. II, London, 1959]

Ans. Interval according to observer A is 5 sec

Interval according to observer C is $\dfrac{10\sqrt{3}}{4+\sqrt{3}}$ sec

B moves a distance $\dfrac{20(3+\sqrt{3})\mathbf{c}}{(4+\sqrt{3})^2}$

with respect to C and a distance $5\sqrt{3}\mathbf{c}/2$ with respect to A.
Distance between A and B according to $C = 15\mathbf{c}/(4+\sqrt{3})$.

Problem 6.10

A 60° glass prism (refractive index 1·5) is given a velocity v, relative to a laboratory, normal to one of the three prism faces. Find the direction in the laboratory of a beam of light which, passing through the other two faces, will have a symmetrical path through the moving prism.

[B.Sc. (Special), Sheffield, 1962]

Ans. $\tan^{-1}\dfrac{2{\cdot}97}{\sqrt{\left(1-\dfrac{v^2}{\mathbf{c}^2}\right)}}$

Problem 6.11

What experimental evidence is there for the special theory of relativity? Starting from the assumption that the velocity of light in free space is constant, deduce the Lorentz transformation connecting the measurements of length and time made by two observers in uniform relative motion.

Two rockets leave the earth and rapidly accelerate to a velocity relative to the earth of one-half the velocity of light. After one year at constant speed, as measured by a clock in the rocket, one rocket reverses its velocity and returns to earth. After ten years at constant speed the second rocket also reverses its velocity and returns to earth. Assuming that the rockets are similar and accelerate and decelerate in a similar way, discuss and obtain a value for the difference in the times, as measured on the earth, spent by the rockets away from the earth.

[Nat. Science Preliminary, Pt. II, Cambridge, 1960]

Ans. 20·78 years

Problem 6.12

Using the fact that the phase of a light wave in free space is invariant under the Lorentz transformation derive an expression for the Doppler shift in frequency of radiation emitted from a source moving with respect to an observer.

A monochromatic beam of light of wavelength 5,400 Å is emitted from a space station and is reflected back by a rocket travelling with uniform velocity directly away from the station. The reflected light is shifted in wavelength by 1,200 Å. Is this shift an increase or decrease in wavelength and what is the velocity of the rocket?

[B.Sc. Higher, Pt. II, III, Liverpool, 1963]

Ans. 0·1c, increase in wavelength

Problem 6.13

The phase of a light wave must be invariant for Lorentz transformations. Use this fact to deduce a relationship between the frequencies determined for a light beam by two different observers moving with a relative velocity v parallel to the direction of propagation of the light beam.

If the source of light and one of the observers are seen by the other observer to be moving in exactly opposite directions with equal velocities 0·6 **c**, and this observer also measures the wavelength of the light as 4,000 Å, calculate the frequency of the source and the wavelength observed by the other observer.

[B.Sc. Higher, Pt. II, III, Liverpool, 1963]

Ans. $1{\cdot}5 \times 10^{-15}$ c/s.
8,000 Å.